THE QUESTION OF CAFFEINE

Edited by **Jolanta Natalia Latosińska**
and **Magdalena Latosińska**

The Question of Caffeine
http://dx.doi.org/10.5772/65133
Edited by Jolanta Natalia Latosińska and Magdalena Latosińska

Contributors

Ana Carolina Luchiari, Luana Santos, Julia Ruiz-Oliveira, Priscila Silva, Patay Eva Brigitta, Antonescu Ina Andreea, Antonescu Angela, Dobjanschi Luciana, Fritea Luminita, Bianca Yadira Pérez-Sariñana, Sergio Saldaña-Trinidad, Tatsuhiko Miyazaki, Keiichi Kato, Hiroshi Nakata, Carla Severini, Wojciech Sledz, Agata Motyka, Sabina Zoledowska, Agnieszka Paczek, Emilia Los, Jacek Rischka, Jolanta Natalia Latosińska

Published by InTech
Janeza Trdine 9, 51000 Rijeka, Croatia

Publishing Process Manager Maja Bozicevic
Technical Editor SPi Global
Cover InTech Design team

First published June, 2017
Printed in Croatia
Legal deposit, Croatia: National and University Library in Zagreb

Additional hard copies can be obtained from orders@intechopen.com

The Question of Caffeine, Edited by Jolanta Natalia Latosińska and Magdalena Latosińska
p. cm.
Print ISBN 978-953-51-3273-8
Online ISBN 978-953-51-3274-5

For EU product safety concerns:
IN TECH d.o.o., Prolaz Marije Krucifikse Kozulić 3, 51000 Rijeka, Croatia,
info@intechopen.com or visit our website at intechopen.com.

Meet the editors

Editor, Dr. Jolanta Natalia Latosińska is a mathematics and physics graduate from the Faculty of Mathematics and Physics, Adam Mickiewicz University in Poznań, Poland. She has got her PhD degree in physics and since 1998 has been working as adjunct at the Faculty of Physics, AMU. She was conferred habilitation degree in physics in 2004. Her fields of interests cover a broad spectrum of topics starting from the studies of the structure-dynamics-activity of drugs by magnetic resonance spectroscopy and quantum chemical calculations to modelling the ultraviolet radiation level using neural networks. All her studies are somehow related to ways of fighting cancer. She published more than 90 SCI journal papers, 5 chapters in monographs, 2 books and 92 conference communications. She was a reviewer of more than 45 journals and a member of the editorial board of 2 SCI-listed journals.

Co-Editor, Magdalena Latosińska, MSc, is a physics graduate from the Faculty of Mathematics and Physics, Adam Mickiewicz University in Poznań, Poland. She is working as a lecturer at the Faculty of Physics, AMU. Her fields of interests cover a broad spectrum of topics starting from the aspects of visualization in the studies of structure-dynamics-activity of drugs to modelling the ultraviolet radiation level using neural networks. She published more than 26 SCI journal papers, 3 chapters in monographs and 17 conference communications.

Contents

Preface

Caffeine-containing food/drinks have been known almost on every continent outside Europe (tea - Eastern Asia; mate and guarana - South America; cocoa - Mesoamerica; coffee and kola nuts - Africa) from ancient times. Europe, till the Enlightenment era, had been drinking alcoholic drinks (beers, wines, distillates and meads) but in the seventeenth century started to drink tea, which came from China, through Japan; coffee, which arrived from Western and Minor Asia and cocoa, which came from Mesoamerica. Europeans discovered that both alcohol and caffeine fuddle the brain, but alcohol acts as a depressant and induces fatigue, while caffeine temporarily restores alertness. This made them so willing to replace everyday alcoholic drinks with caffeinated ones. But with stimulating coffee, a spirit of revolution, which broke with feudal politics and opened up the epoch of modern politics and industry, overtook Europe, initially France and later the United States. Modern young societies demanded more and more caffeine supplied in different forms by the overseas colonies, which did not keep pace with market needs. Thus, surprisingly shortly after the isolation of caffeine from coffee beans and its identification as an active component, pure caffeine has been found extremely attractive. Its use in various pharmacological applications as a stand-alone drug and an additive to other drugs *(analgesic, anti-migraine and cardiac)* made it gradually the most popular psychoactive substance in the world and started a new era in pharmaceutics. Furthermore, modern chemical industry went one step further with the use of pure caffeine discovery. Soft drinks, energy drinks and shots containing pure anhydrous caffeine (like North American Coca-Cola, European Lucozade, Japanese Lipovitan or Austrian Red Bull) have become widely available in each shop, gas station or drinks machine, simply flooded the markets in the whole world. Caffeine-driven world has accelerated and still accelerates.

Because of its ability to reduce tiredness, sleep deprivation and improve alertness, caffeine emerged in the twenty-first century as a miraculous specific, which allows humans to cross their normal physiological and psychological body limits. In fact, caffeine is a booster fooling the brain and seemingly *fighting physical fatigue*. Caffeine attractiveness comes from its natural i.e. 'healthy' origins and strong psycho-stimulating properties, with relatively weak side effects. Uncounted publications, reviews, studies, theses and books have been devoted to caffeine—its form, structure, metabolism and positive/negative influence on practically each part of the human body. However, with its new study, more and more ambiguities appeared.

Paracelsus prophetically wrote in the fifteenth century that 'Poison is in everything, and no thing is without poison. The dosage makes it either a poison or a remedy'. But it is difficult to predict safe amount of caffeine, which can be consumed without any consequences. Why? Because each organism reacts differently. Caffeine case proves that the matter is not only the dose but also our inherited genotype. The ability to identify associations between inherited genotype, disease and drugs still requires appropriate epidemiological, clinical and pharmaceutical studies. But caffeine studies carry the hope to understand these associations and find better targeted, more efficient and highly personalized, treatments for various diseases. Its ability to pass easily through the blood-brain barrier and improve blood circulation can be helpful to develop more sophisticated drug delivery methods, especially those aiming at the brain or bones. Natural insecticidal and repellent properties of caffeine make it a convenient starting point to search for new, safer methods/ways of protecting plants (crops, greenhouse cultivations and overwinter storages). But caffeine, which is stimulating for the human body, also delivers a cheap, renewable fuel.

This book consists of chapters covering caffeine history, methods of its determination and not only astonishing medicinal but also non-medicinal applications. It is our hope that every reader will find in this book something interesting, inspiring, informative and stimulating.

Jolanta Natalia Latosińska and **Magdalena Latosińska**
Faculty of Physics, Adam Mickiewicz University,
Poznań, Poland

Chapter 1

Introductory Chapter: Caffeine, a Major Component of Nectar of the Gods and Favourite Beverage of Kings, Popes, Artists and Revolutionists, a Drug or a Poison?

Magdalena Latosińska and
Jolanta Natalia Latosińska

Additional information is available at the end of the chapter

http://dx.doi.org/10.5772/intechopen.69693

1. Caffeine consumption around the World

Global caffeine consumption is estimated to be around 120,000 tonnes per year, which corresponded to one cup of coffee per day for every human on the planet. Based on the statistics, the top tea-producing countries in the world are: China, India, Kenya, Sri Lanka, Turkey, Indonesia, Vietnam, Japan, Iran and Argentina. Main producers and exporters of coffee are: Brazil, Vietnam, Colombia, Indonesia, Ethiopia, India, Honduras, Uganda, Mexico and Guatemala, **Figure 1**.

Caffeine consumption is the highest in tonnes in the United States (971), followed by Brazil (969), Germany (425), Italy (211) and France (202). About 79% of total consumed caffeine comes from coffee, 15% from tea, only 3% from mate and 4% from cocoa [1]. In general, people in the west drink more coffee, while the eastern world drinks more tea, **Figure 1**. Tea consumption per capita predominates in Turkey, Russia, Iran, Mauritania, Syria and China. In Paraguay, Argentina and Brazil, the consumption of mate is dominant. The rest of the world prefers coffee. Europeans are the world's biggest coffee drinkers. Coffee consumption in Europe varies from around 10 kg per capita per year in the Nordic countries (Finland, Norway) to around 3 kg per capita per year in the United Kingdom and most Eastern Europe countries. Annual consumption over 5 kg per capita per year in Brazil is exceptionally high among over 60 coffee exporters. The largest cocoa consumption is noted in Switzerland, Germany, Ireland, the United Kingdom and Norway. The world's biggest Coca-Cola drinkers are in Mexico, Chile, the United States, Panama and Argentina. Energy drinks containing caffeine like Red Bull, Monster, Suntory, Rockstar have experienced a considerable growth in popularity in the last 25 years, but still represents only 1% of the overall non-alcoholic beverages market. Austria led the global per capita consumption and is followed by Ireland, the United Kingdom, Switzerland, the United States and Australia. Caffeine, in any form, is consumed

Figure 1. Coffee and tea consumption/production around the world in 2015 (statistical data: Food and Agriculture Organization of the United Nations and Euromonitor).

daily by about 90% of adults, which makes this psychoactive, but legal substance the world's most widely used drug.

Despite caffeine huge popularity and its countless studies, there is still much confusion, inconsistencies and contradictions in the results, poorly known side effects and unknown applications.

2. Historical aspects

Caffeine-containing species from *Camellia, Coffea, Cola, Ilex, Paullinia, Theobroma* and *Citrus* genus have been known from ancient times, but phytogenic studies indicated that they are not closely related [2]. However, most of them grow in tropical or sub-tropical zones. *Camellia* originates from Asia, *Coffea* and *Cola* from Africa, *Ilex, Paullinia* and *Theobroma* from America, *Citrus* from Australia and Oceania. Tea (*Camellia sinensis* (L.) Kuntze), coffee (*Coffea arabica* (L.), *Coffea canephora var. Robusta*, (Pierre ex A. Froehner), *Coffea liberica* (Bull ex Hiern)), cacao (*Theobroma cacao* (L.)) and citruses (*Poncirus* (L.) Raf., *Fortunella* (Swingle), *Microcitrus* (Swingle)) have been used as medicinal products, stimulating food, dietary supplement or fragrant plants, while cola (*Cola nitida* (*Vent.*) Shott and Endl.)), mate (*Ilex paraguariensis* (A.St.-Hil.)) and guaraná

http://dx.doi.org/10.5772/intechopen.69693

(*Paullinia guarana* (Kunth) or *Paullinia cupana* (Kunth), *Paullinia sorbilis*, (Mart)) as ritual plants. Almost each country has got its own legends on finding natural source of caffeine.

According to Cha Jing by Lu Yu, a mythical ruler of prehistoric China Shen Nong (also known as Wugushen or Wuguxiandi), reigning 3000 BC, discovered tea, when a few leaves of the nearby tree *Camellia sinensis* (L.) fell into the boiling water [3]. In the times of the Chinese Shang dynasty, tea was used as a medicinal drink, but later, during the Chinese Tang dynasty, it was popularized in East Asia as a recreational drink [4]. The etymology of the word tea goes back to the Chinese 茶(*tê, chá* and *chai)*, which also indicates the region of its origin. The first unambiguous reference to tea treated as a beverage is dated to 59 BC (Western Han dynasty era) [5]. In 805 AD, the seeds of tea were brought to Japan by the Buddhist saint Saichō (Dengyo Daishi). Soon after that, the cultivation of tea in the five provinces surrounding the capital of the country, Kyoto, was ordered by the enthusiastic 52nd Japan emperor Saga. Exactly who first brought tea to Europe in the seventeenth century remains a mystery, but it is known, that the oriental goods including tea have been imported by the Portuguese since 1517 and by the Dutch since 1610. The seventeenth-century apothecaries added tea to other luxury items like sugar, ginger and spices and sold them next to the medicines. In 1658, Katherine Braganza, Portuguese wife of Charles II Stuart, brought tea to England [6]. It is known that French ruler Louis XIV (the Sun King) drunk tea for health reasons starting from 1665 [7]. By 1675, tea was in general use throughout Holland and started to being sold in grocery stores. To Russia, tea was brought from China as a gift to Russian tsars. For the first time, about 1630—it was a gift to Russian tsar Michael I (Romanov) from a Mongol Khan Sholoi [8] and for the second time, in 1680—it was presented to tsar Alexis I from the Chinese ambassador to Moscow [9]. European tea merchants of eighteenth century recognized only three growing markets: Holland, England and Russia. But the fourth one was the young market in British American colony. The Tea Act, legislative manoeuvre by Lord North, passed by the Parliament of the United Kingdom on 10 May 1773, granted the British East India Company Tea a monopoly on tea sales [10]. On 16 December 1773, the Patriot group 'Sons of Liberty' destroyed a shipment of tea in Boston Harbour. This event that became known as the Boston Tea Party was the signal to American War of Independence [10]. Almost 100 years later, in mid-1800, tea was successfully harvested in South Carolina. Although *Camellia sinensis* (L.) originates from East Asia, the Indian Subcontinent and Southeast Asia, but nowadays, it is cultivated in most tropical and subtropical regions of the world.

The history of coffee has its beginnings in the sixth-century Ethopia [11], however, Ethiopian Galla tribe ground up coffee beans (actually the pit of the berry), mixed them with animal fat and consumed as an energy food, much earlier. The famous legend attributes it to the shepherd of Caldas from Abyssinia, who in 525 AD noticed that the goats that had grazed among the bushes became excited and sleep-deprived. After sampling the fruit from the bushes growing there, he experienced a similar surge of strength. Arab traders brought coffee to Yemen [12]. The oldest written references to coffee (*'bunchum'*) were found in Kitab al-Hawi—a comprehensive book on medicine by Abu Bakr Muhammad ibn Zakarijja ar-Razi (the ninth-century Persian polymath, physician, alchemist and philosopher) [13]. By 1414, coffee was known in Mecca and spread to Egypt from Al Mucha (Mocha), the Yemeni port, then to Syria and Istanbul, the capital of the vast Ottoman Turkish Empire [14]. The first coffee shop, Kiva Han, was opened in Constantinople in 1475. In the fifteenth century, the Sufis of Yemen routinely used coffee to stay awake during prayers. There was an attempt to ban coffee in 1511 in Mecca, because religious leaders accounted it for stimulation of the radical thinking, but sultan of Cairo overruled the idea and the ban was lifted. By 1630, over one thousand coffee houses were operated in Cairo. In the end of the sixteenth century, coffee spread throughout the middle East. Coffee arrived to Europe by two routes—from the Ottoman Empire, and by sea—from the original coffee port of Mocha. The German botanic Léonard Rauwolf for the first time described coffee in 1576 in *Viertes Kreutterbuech—darein vil schoene und frembde Kreutter* [15]. In the seventeenth century, coffee was known in Europe as 'Arabian wine' or 'Muslim drink' and thus unpopular. Coffee enthusiast pope Clement VIII 'baptized' it around 1600 [14]. The coffee name comes from the original Arabic *quhwah* through Turkish form *kahveh* translated to Italian as *caffè* or *Danish* as *kaffe*. Shortly after the first 'cafes' in Venice, Oxford, London were established. When Turkish siege of Vienna in 1683 was broken, the European victor Johan III Sobieski allowed Jerzy Franciszek Kulczycki, sas coat of arms, to choose as a reward anything from the Turkish camp. Amazingly, Kulczycki opted for 300 bags containing the 'strange seed' (huge coffee supplies). The legend says that Kulczycki opened the first coffee house *Hof zur Blauen Flasche* in Vienna in 1683 [16]. Cafes quickly gained popularity throughout the whole western Europe playing a significant role in shaping social relations. In 1650, Jacobs, a Lebanese Jew, opened the first coffee house in Oxford, England [11]. Shortly thereafter, cafes where people could buy coffee for 1 penny and carry on intellectual conversations, called 'penny universities', began to emerge. Famous *Café Procope* in Paris, a gathering place of many French notables, actors, writers, philosophers and musicians, was opened in 1689 by Francois Procope, a Sicilian who came from Florence [17]. The parts of the furniture of this café were Voltaire, Denis Diderot, Pierre Beaumarchais, Honoré Balzac, Victor Hugo, Paul Verlaine, Jean-Jacques Rousseau, fathers of French revolution: Jean-Paul Marat, Maximilien de Robespierre, Georges Danton and young Napoleon Bonaparte, later France emperor. By 1843, the number of cafes in Paris increased to as many as 3000. The first coffee houses in Germany were opened in Regensburg and Leipzig. Johann Sebastian Bach, Leipziger, the most heavy coffee drinker ever, wrote the Coffee Cantata in its honour. The first and the oldest to date café in Salzburg was *Café Tomaselli* founded in 1700. Frequent cafes guests were Wolfgang Amadeus Mozart, Michael Haydn, Hugo von Hofmannsthal and Max Reinhardt. In Russia, historically, the tradition of coffee-drinking was introduced by Peter the Great, who brought it from his travel to the Netherlands [18] and was fostered by Empress Catherine II the Great [19]. It must be said that in those days not all were coffee lovers. King Frederick II of Prussia even issued a manifesto claiming beer's superiority over coffee and charged a heavy tax on coffee commercialization in 1777 [20]. Coffee reached New Amsterdam (New York) in mid-seventeenth century and then the New World. It immediately obtained a status of one of the most popular drinks. As the demand steadily grew, there was strong competition to cultivate coffee outside of Arabia. The first attempts to plant coffee by the Dutch failed in India, but were successful in Indonesia (Java, Sumatra and Celebes). In 1714, a young coffee plant was given by Gerrit Hooft, the Mayor of Amsterdam, to King Louis XIV of France as a gift [14]. It was carefully planted in the Royal Botanical Garden in Paris. Nine years later, a seedling stolen from this plantation by king's doctor was transported to Martinique by Gabriel de Clieu [14]. It was the nucleus of about 18 million trees plantation in Martinique 50 years later. This seedling was also a parent of all the coffee trees throughout the Caribbean, South and Central America. The Brazilian coffee trees also come from France, exactly from French Guiana. Francisco de Mello Palheta was a military responsible for the introduction of coffee cultivation in Brazil [21]. Despite many attempts, he was not able to get coffee plants officially, but in 1720, Marie-Claude de Vicq de Pontgibaud, the wife of the French governor Guiana Claude Guillouet d'Orvilliers, smashed the handful of seeds inside the bouquet of flowers—a farewell gift. Quickly the cultivation of coffee had been introduced in Dutch Guiana (1714), Jamaica (1718) and expanded to the tropical regions of South America. Throughout the nineteenth and the first decades of the twentieth century, Brazil was a monopolist on the coffee market, but later, Colombia, Guatemala and Indonesia started to cultivate coffee. European colonial regimes initiated the coffee cultivation and export in Kenya, Angola, Uganda and Ethiopia, where it all started. During the American Civil War (1861–1865), Union soldiers received from the government in Washington 36 pounds of coffee annually (about 16.3 kg), because without coffee soldiers did not exist. The status of coffee had changed from the scarce elixir into a public beverage.

Another source of caffeine is cocoa (*Theobroma cacao* (L.)) nuts [22], used by pre-Olmec cultures in Mexico as early as 1900 BC. Olmec, Mayan, Toltec and Aztec civilizations used chocolate as an invigorating drink, stimulating mystical and spiritual qualities [23]. In the New World, chocolate was consumed in the form of a bitter and sharp drink called xocoatl, containing a bit of vanilla, chilli peppers and achiote. Cocoa seeds in pre-Columbian Mesoamerica were luxury goods and used as a means of payment (currency). In 1517, the Spanish conquistador Don Hernán Cortés [23] was treated to xocoatl by the Aztec emperor Montezuma. Eleven years later, he brought xocoatl to Spain, where it became a popular drink on the Spanish royal court. The name of this drink comes from the Nahuatl words *xocoatl* (xoco 'bitter' and atl 'water') and *cacahuatl* (cacao) translated to Spanish as *chocolate*. Spain kept chocolate secret for nearly a century, but when Anne of Austria, the daughter of Spanish king Philip III wed the French king Louis XIII in 1615 [24], chocolate spread across Europe. In 1689, Hans Sloane invented a sweet milky version of this drink, which was originally prepared by local apothecaries until 1897, when the Cadbury brothers acquired the exclusive right to manufacture it [25]. As demand for cocoa increased, its plantations were established in the West Indies (Caribbean Basin), Philippines, Asia and Africa. Due to the technological improvement—cocoa press—invented by the Dutch Casparus van Houten Sr., chocolate-making process was revolutionized [26]. Since then pulverization of cocoa into cocoa powder became a basic step in production of all chocolate products. In 1847, British company J.S. Fry & Sons produced first chocolate bar using cocoa butter, cocoa powder and sugar [27]. Shortly after that bars of chocolate flood the whole Europe. In 1879, in Berne, Switzerland, Rodolphe Lindt invented the conching machine, which gives chocolate a velvety texture and superior taste [26]. A chocolate boom which started in the late 1800s and early 1900s still has not slowed down. During the Second World War, bars of chocolate were the emergency store of each Swiss or US army soldier.

Kola (*Cola acuminata* and *Cola nitida Schott & Endl.)* [28], a tree native to the tropical Africa known from at least the fourteenth century is a natural source of caffeine. The etymology of the word *kola* derives it from the Latinized form of a West African name of the tree. The kola nuts were chewed in many West African cultures to restore vitality and as appetite suppressant able to alleviate the feeling of hunger [29]. African exports to England and the United States started only in the mid-nineteenth century. The worldwide career of kola began in 1886 when John Pemberton from Atlanta, Georgia, created a recipe of 'Coca-Cola' [30]—an extract based on mixed kola and cocaine, used as a headache and hangover remedy [31].

Another old, but much less popular source of caffeine are the leaves and stalks of three species of holly tree genus *Ilex vomitoria* (Sol. ex Aiton) *(Saint Yaupon)*, *Ilex paraguariensis* (A.St.-Hil.) *(Yerba Mate)* and *Ilex guayusa* (Loes.). *Ilex vomitoria* (Sol. ex Aiton) has been used by the North American Indians to brew tea called Asi (black tea) from the archaic era. It contains up to six times more caffeine than strong coffee and provokes vomiting for cleansing the body and soul. In South America (Argentina, Uruguay and Paraguay), a drink called yerba mate was made of *Ilex paraguariensis* (A.St.-Hil.). The Brazilian name is Chimarrão (Erva Mate chimarrão). Yerba Mate name comes from the Spanish *yerba* and *mati*, which in Quechua means *gourd*. Legend tells that when Yarí, the moon and Araí, the pinkish cloud of dusk, came to visit the Earth, a jaguar attacked them. They were rescued by an old Indian, who received in a reward this new kind of plant. People of the indigenous cultures in Argentina, Brazil, Paraguay and Uruguay who have survived periods of drought by drinking yerba mate called it 'Drink of the Gods'. This source of natural caffeine was popularized in Europe as an alternative to Asian tea by Jesuit missionaries who arrived to the Parish basin in the mid-seventeenth century and appreciated the advantages of a beverage made from powdered leaves and shoots. *Ilex guayusa* (Loes.) Amazon tree comes from tropical rainforest of Ecuador but is grown in Peru and Columbia. It is a completely unpopular, but rich source of caffeine, similar to coffee. In contradiction to the other caffeine containing beverages, drink made of its leaves is not only stimulant but also energizing, relaxing, calming and can cause conscious dreams. A great lover of yerba mate is pope Francis, native Argentinean. An exclusive drinking yerba mate kit was a present for pope Francis from the Argentine President Cristina Fernandez de Kirchner during her first audience in Vatican. Che Guevara, Lula da Silva, Jorge Luis Borges, Julio Cortázar, Barack Obama, Hillary Clinton and Madonna are all well-known yerba mate drinkers.

Also guaraná (*Paullinia guarana* (Kunth), *Paullinia cupana* (Kunth) *and Paullinia sorbilis* (Mart)) seeds, named after the Guarani Indian tribes, have been used for centuries by the inhabitants of the Amazon basin to restore lost forces. In the early eighteenth century, guaraná has been discovered and classified by the German botanist C.F. Paulini. Commercial use of guaraná began to spread after 1958, because it became an indispensable ingredient in many brewed beverages produced in Brazil and the United States.

Citrus (L.) (all true citrus trees including *Poncirus* (L.), *Fortunella* (Swingle) and *Microcitrus* (Swingle)), the weakest source of caffeine, originates from Australia, New Caledonia, New Guinea [32] and probably Southeast Asia bordered by India, Myanmar and China. The etymology of the word *citrus* derives it from the genus name in modern Latin. Although *Citrus* species leaves and flowers contain caffeine [33], they have been cultivated since ancient times mainly for fruits, in which caffeine is not present. However, citron leaves in sugar or honey or Korean honey citron tea (Yuja Cha) made of boiled leaves have also been highly popularized [34]. The fragrances, flavours and oils made of citrus have been known and desirable for many centuries in medicine and perfumery. The oldest traces of citrus in Europe date back to thirteenth-century BC Cyprus. The earliest fragrances (e.g. Eau de *Cologne* 1709 by Farina, Imperial 1850 and Eau de Imperiale 1861 by Guerlain, Jicky 1889 by Coty) contained bergamot, lemon, lime, mandarin and orange blossom oil [35]. Since then the popularity of citrus-spirit type of perfume or eau de toilette has not decreased. Small quantities of caffeine contain some types of honey (e.g. Greek orange honey), because citrus and coffee plants attract bees using caffeine as a part of rewarding system [36, 37].

3. Health considerations

For a long time, it has been a dilemma if coffee and tea are non-toxic and which is better for health—tea or coffee. From among all natural sources of caffeine, only tea started a career as a medicine and became a beverage in the course of time. In the eighteenth century, the Swedish king Gustav III, proposed the twin brothers who were sentenced to death for murder, a death row pardon in exchange for their participation in the scientific experiment [38]. One of the twins had to drink four cups of coffee a day, the other four cups of tea a day. A group of professors from Swedish Kings Academy of Sciences examined them to check the influence of these beverages on their organisms. The twins drank and drank, in the meantime, the king was murdered, the professors died. The first died the tea-drinking brother, while the compulsory coffee fun lived several years longer. But the tea drinker died at the age of 84, which at the time when the average life span was about 40, was considered as unbelievable achievement. What about the final verdict? No doubt by this simple long-lasting experiment, both dietary habits were considered as an important factor positively influencing human health. But the question remained which turned out better for health, tea or coffee, and first of all what factors were responsible for it.

Although all these natural sources of caffeine have been used for a long time as a beverage or drug, the fact that caffeine is the main factor responsible for their effect remained a mystery. Only in 1819, at the personal request of Johann Wolfgang von Goethe, the relatively pure chemical form of caffeine was isolated by Friedrich Ferdinand Runge [11], who called it '*Kaffebase*'. Eight years later, in 1827, M. Oudry obtained '*theine*' from tea [39]. In 1838, Mulder [40] and Jobst [41] showed that theine was actually caffeine. Thus, taking into account caffeine input both tea and coffee should be similarly health-promoting, which would not be a surprising result today, as we know main chemical component. The molecular structure of caffeine (1,3,7-trimethylxanthine; 1,3,7-trimethyl-1*H*-purine-2,6-(3*H*,7*H*)-dione) was described in 1882 by Hermann Emil Fischer, who also made its first complete synthesis, for which he was awarded the 1902 Nobel Prize [42]. He showed that caffeine found in coffee is equivalent to those in tea and cacao. Nowadays, caffeine is still rarely obtained by total chemical synthesis or semi-synthetic processes, which are economically inefficient. Instead, it is extracted from plants often as a by-product in the manufacture of decaffeinated coffee, **Table 1**.

Caffeine source	Origin	Plant	Plant part	Caffeine concentration per milligram (%)	No. of all chemical compounds
Tea	natural	*Camellia sinensis* (L.)	Leaf or shoot	4.8–9.3*	771
Coffee	natural	*Coffea arabica* (L.)	Bean or fruit	0.06–3.2*	154
Cacao	natural	*Theobroma cacao* (L.)	Seed	0.062–1.29*	261
Mate	natural	*Ilex paraguariensis* (A.St.-Hil.)	Leaf	0.2–2.0*	39
Guarana	natural	*Paullinia cupana* (Kunth.)	Seed or fruit	0.9–7.6*	23
Kola	natural	*Cola acuminata* (Schott & Endl.)	Seed	1.5–2.5*	9
Citrus	natural	*Poncirus* (L.), *Fortunella (Swingle)*, *Microcitrus* (Swingle)	Leaf or flower	0–0.008*	495
Caffeine anhydrous	synthetic	-	-	>98.5	1
Dicaffeine malate	synthetic	-	-	65–70	2
Caffeine citrate	synthetic	-	-	45–55	3

***from** Dr. Duke's Phytochemical and Ethnobotanical Databases (https://phytochem.nal.usda.gov/).

Table 1. Naturally occurring in plants and synthetic caffeine doses.

When it seemed that everything was known about the structure of caffeine, it turned out that the matter was much more complicated—an untypical polymorphism of caffeine was discovered [43]. An anhydrous caffeine exists in two enatiotropically related polymorphic forms: stable (phase II or β-form) which melts at 508K and metastable (phase I or α-form) melting at 512K [44] and each form displays different physicochemical properties [45].

Some authors consider the existence of phase III [46], while the others a mixture of two phases I and II [47]. The phenomenon of polymorphism further complicates the co-existence of structural and dynamical disorder. A number of experimental techniques (e.g. X-ray [47–49], synchrotron X-ray diffraction [50] mid-infrared (MIR), near-infrared (NIR) Raman spectroscopies [51, 52], dielectric measurements [46], NMR-NQR spectroscopy [53, 54]) have been applied to clarify the matter but still new doubts arise. Screening of polymorphs is of importance due to the differences in solubility, long-term stability, dissolution rate and bioavailability. Many novel beverages like soda or energy drinks [55] as well as drugs contain pure caffeine, thus there is considerable public health interest in its effects on humans.

Because caffeine is the most widely used stimulant, its metabolism and effect on the human body have been intensively studied. Caffeine is known to stimulate the central nervous system (affects sleep, arousal, cognition, learning and memory), as well as muscular, respiratory and circular systems [56–59]. But it is supposed that a broad spectrum of caffeine effects is a result of action of its metabolites. Caffeine demethylation yields to about 4–5.4% of theophylline, 10.8–12% of theobromine and 81.5–84% of paraxanthine [60]. While caffeine, theophylline and theobromine naturally occur in about 80 green plants species, paraxanthine does not, because it is not accumulated in plants due to the very slow N1-methylation of 7-methylxanthine [61, 62]. But, paraxanthine discovered in human urine by Solmon [61] results from demethylation of caffeine at the 3-position through the catalytic action of polymorphic cytochrome P450 subtypes 1A2 (90%) and 1A1, 2E1, 3A4 and 2D6 (10%) [63, 64]. It was discovered that caffeine and its metabolites belong to the pharmacological group of adenosine A-receptor (A_1, A_{2A}, A_{2B} and A_3) antagonists [65]. The A_1 and A_2 receptors bind caffeine at low doses and the A_{2B} receptor at high doses. The A_3 is caffeine insensitive. Caffeine and its metabolites theophylline and theobromine act primarily as non-selective antagonists at A_1 and A_{2A} receptors in both human central nervous system and heart. Surprisingly, only paraxanthine acts similarly to caffeine [66], theobromine acts as vasodilator, diuretic and heart stimulant [67], theophylline relaxes smooth muscles of the bronchi and is effective in chronic obstructive pulmonary disease and asthma [68]. Theobromine is a weaker antagonist of adenosine receptors and therefore has a lesser impact on central nervous system, but stronger on heart. Most caffeine activity has been attributed to this antagonism and raised attention to it as potential parent compound in designing dual-target-directed drugs that simultaneously inhibit monoamine oxidase B (MAO-B) and antagonize adenosine A_{2A} receptors ($AA_{2A}R$) in the brain [69]. But caffeine also acts by the inhibition of non-adenosine receptor $GABA_A$, an ionotropic receptor, responsible for most of the physiological activities of GABA in the central nervous system [70], while paraxanthine by the inhibition of cyclic guanosine monophosphate (cGMP), which is a key-factor for anti-inflammatory and psychostimulant effects [71].

It is known that caffeine has the ability to reduce the physical, cellular and molecular damage caused by spinal cord injury (SCI), stroke or neurodegenerative chronic diseases of Parkinson [72–74] and Alzheimer's [75–78]. But it has been reported that paraxanthine, rather than caffeine itself, reduces the risk of developing Parkinson's disease [79, 80] and contrary to caffeine it is strongly protective against neurodegeneration and loss of synaptic function [71]. Besides, caffeine exhibits inhibitory activity against diabetes II, gallstones and cirrhosis of the liver [81]. It acts as diuretic [82, 83] and stimulate tear secretion [84] which makes it helpful in the dry eye syndrome treatment [85]. Antioxidant properties of caffeine and scavenging abilities of reactive oxygen species (ROS) are associated with its ability to reduce the risk of liver, kidney, basal, colorectal and endometrial cancers [86–90]. Only recently caffeine-based gold compound has been discovered as a potential anti-cancer drug selective for ovarian cancer [91]. Caffeine mitigates the adverse mutagenic effect of ultraviolet radiation [92–95] or anti-cancer drugs [96–98]. It is difficult to study pure caffeine effect on health because it is consumed with many different additional chemical compounds (tea up to 771, coffee up 154, cacao up to 261, mate up to 39, guarana up to 23, kola up to 9 and citrus up to 495), **Table 1**. The problem is further complicated by the presence of metabolites of caffeine in their composition.

Such a broad spectrum of its action has stimulated a significant interest in studies of caffeine at much more sophisticated level, which should explain the differences in the individual reactions to caffeine. How we react to caffeine varies between individuals because it is largely dependent on individual genome. The earliest studies on the possible link between genes and coffee consumption date back to the 1960s [99]. Although a number of further twin experiments provided some evidence for the heritability factor in response to caffeine [100], the genetic contribution to caffeine consumption strongly depends on sex and decreases with age. Thus, true importance of individual genetic variability has been testified in larger diverse populations and focused on caffeine rich diet-disease studies at molecular level [101, 102]. According to them, five genes *CYP1A2, AHR, ADORA2A, COMT* and *PDSS2* are known to be related to the caffeine sensitivity. Gene '*CYP1A2*' releases the liver CYP1A2 enzyme, which breaks down caffeine [103, 104]. 'COMT' controls the breakdown of catecholamines, 'AHR' controls the state on/off of CYP1A2 [105, 106], 'PDSS2' regulates the production of CYP1A2 [80] and 'ADORA2A' is responsible for the variation of A2A to which caffeine binds and controls caffeine sensitivity [107].

Although coffee intake has been supposed to be a risk factor for heart disease, it was not related to genes. The enzyme catechol-O-methyl transferase (COMT) is known to break down catecholamines, which in high concentrations can induce a heart attack. Due to variability of the '*COMT*' gene, the COMT enzyme has a number of variants. For example, the COMT rs4680 variant is accompanied with low level of COMT enzyme. But in the presence of caffeine, the release of catecholamines strongly increases and thus a risk of a heart attack also increases [108]. The gene '*CYP1A2*' releases the key enzyme that breaks down caffeine. Two variations of this gene (*CYP1A2*1A*—high activity and *CYP1A2*1F*—low activity, differing in nucleotide and marked by A->C substitution at position 734) help metabolize caffeine: one faster and the other one slower [103, 104, 109]. Because every person has two copies of this gene inherited from each parent, a particular combination is responsible for the speed of one's own metabolism (fast in the case of fast + fast, and slow in the case of fast + slow/slow + slow combinations) [103, 104, 110]. Fast metabolism is of course beneficial as it is related to much (22%) lower risk of heart attack and higher fertility. But CYP1A2 is also a key enzyme in the activation of carcinogenic heterocyclic aromatic compounds [101]. Thus, caffeine consumption has been associated with ovarian cancer risk, which strongly depends on the variations in *CYP1A2* genotype (high-inducibility A/A and low C). A similar study has shown that caffeine consumption protects only women with a BRCA1 mutation against breast cancer [102]. A genome-wide association study on two populations in Italy and the Netherland allowed identification of a *PDSS2* gene that regulates the production of proteins metabolizing caffeine in the human body. The higher levels of this gene result in a slower caffeine metabolism and necessity to drink less amounts of coffee [80]. It has been found that a common variation in *ADORAA2A* is also associated with caffeine sensitivity. Two copies of C allele of *ADORA2A* induce sleep disturbances caused by intake of caffeine [107, 105] while two copies of the T allele of *ADORA2A* result in an increase of anxiety level after caffeine [104]. These observations are helpful in explanation of the habitual coffee consumption [110] as well as in the understanding of differences in the individual reaction to caffeine. Although not all caffeine consumers suffer caffeine withdrawal symptoms, but it is so common that in 2013, it was added by the American Psychiatric Association to the Diagnostic and Statistical Manual of Mental Disorders. A particular combination of the variants of five genes mentioned above may significantly increase or decrease a risk of disease or poor tolerance. Thus, the intake of caffeine can have both positive and negative health effects. The International Agency for Research on Cancer only recently, in 2016, revised its classification from 1991 and moved coffee from Group 2b ('Possibly carcinogenic to humans') to Group 3 ('Not classifiable as to carcinogenicity'). This category is used for compounds for which the statistical evidence of carcinogenicity is inadequate in humans or limited in experimental animals. But that does not mean that its safety is not deceptive. It just indicates explicitly that our knowledge is still incomplete.

http://dx.doi.org/10.5772/intechopen.69693

One more aspect related to the individual caffeine sensitivity should be mentioned—the difficulties in estimation of caffeine lethal dose (LD50), which is about 150–200 mg/kg [111, 112] i.e. 80–100 cups of coffee. When we compare a case of death after ingestion of 6.5 g/person and a case of survival after ingestion of 24 g/person [113, 114], the range of tolerance/intolerance makes an impression and is a warning. Too much caffeine in a few cans of energy drink had killed a 19-year-old Austrian football player, 33-year-old Brooklyn construction's worker or three Swedish teenagers. The statistical data of victims of caffeine overdosing collected by National Poison Data System in the United States indicate that 67% of all 6309 cases of poisoning affect children and adolescents under 20. How much caffeine was in the caps of coffee which Honoré de Balzac, true coffee lover, drank in 60 coffee cups per/day? Caffeine content in popular drinks is collected in **Table 2**. The US Food and Drug Administration,

	Caffeine drink	Size in oz (ml)	Caffeine (mg)
Coffee	Brewed	8 (237)	95–165
	Brewed, decaffeinated	8 (237)	2–5
	Espresso	1 (30)	47–64
	Espresso, decaffeinated	1 (30)	0
	Instant	8 (237)	63
	Instant, decaffeinated	8 (237)	2
	Latte or mocha	8 (237)	63–126
Tea	Brewed black	8 (237)	25–48
	Brewed black, decaffeinated	8 (237)	2–5
	Brewed green	8 (237)	25–29
	Instant	8 (237)	40
	Ready-to-drink, bottled	8 (237)	5–40
	Green tea	8 (237)	25
	White tea	8 (237)	28
	Yerba mate	8 (237)	85
	Guayusa	8 (237)	66
Soda	Coca Cola	8 (237)	24–46
	Pepsi Cola	8 (237)	25
Energy drinks	Energy drink	8 (237)	27–164
	Energy shot	1 (30)	40–100
Shots	Liquid caffeine	1 (30)	500
	NoDoz	1.89 (56)	115
Chemicals	Pure anhydrous caffeine	1 teaspoon (5 g)	4706

Table 2. Caffeine content in popular drinks.

FDA, recently issued warnings due to risk to consumers for overdosing caffeinated products containing pure powdered caffeine. A single teaspoon of pure anhydrous caffeine (5 g) is roughly equivalent to the amount in 28 cups of brewed coffee or in 6 energy shots, **Table 2**.

Similarly to humans, the individual sensitivity and additionally breed/division diversity have also been observed in animals. A poor ability to metabolize caffeine which makes it toxic to dogs, cats [115–118] and birds [119, 120] is quite well documented in domestic animals. The toxic doses are so small that single chocolate bar can kill our beloved pet. But 'Creme Puff' cat, the 'oldest cat ever', listed in the *Guinness Book of World Records* for living 38 years, was served coffee with cream every day by its owner [121]. Caffeine is also known to be harmful to wild organisms like molluscs [122], insects [123] and spiders [124], thus making a part of the natural defence of plants against herbivores, larvae of mealworms, mosquitoes [123], tobacco hornworms, snugs and snails [122]. However, there are awesome exceptions like coffee berry borer, which can reduce a crop yield by 80% and survive the dose equal to 500 shots of espresso/person [125].

4. Final remarks

Caffeine is a chemical component of the oldest known food plants (about 5000 years), the most widely consumed (not counting water) and the most extensively studied (1,468 books, 39,551 journal articles, 2,211 dissertations) component of diet. The seeds or seedlings of plants containing caffeine were stolen, smuggled, hated and desired, accused of demonic or radical influence—banned and baptized. The wars for plantations/colonies were fought and fortunes gained and lost. Caffeine drinks were used in religious asceticism and creative amok, behind closed doors of the cafes were written operas, manifestos and revolutions started. After all coffee seeds were used as a currency and reward, tea and chocolate were sipped by emperors, kings and tsars, coffee was loved by artists, writers, musicians, philosophers, students, popes, revolutionists and belt down by soldiers, mate is preferred by actual pope, a few presidents, writers and celebrities, and energy-drinks containing pure caffeine are nowadays trendy and desired by teens and adolescents. Caffeine under the pretence of tea or coffee changed social manners and war results—coffee has been considered the '*soldiers drink*' since Napoleon. Energy drinks like Coca-Cola, Pepsi, Dr. Enuf, Power Horse or Red Bull containing large amount of pure caffeine fight physical fatigue, increase vigilance and reaction speed and allow people to function almost without sleep, but sometimes they are deadly.

Day by day we are coming into contact with caffeine—in drinks (coffee, tea, soft-drinks as Coca-Cola, soda, chocolate, energy drinks), drugs (above 50 different drugs contain *Coffeinum, Coffeinum Natrium benzoicum, Coffeinum Natrium salicylicum, Coffeinum citricum, Phenazonum Coffeinum citricum*), cosmetics and personal care products, bath (e.g. giant caffeine spa in Japan), even fuels (e.g. 'car-puccino'). We deliver it to inside and outside of our bodies in large amounts but as a matter of fact, we still do not know much about it, because it jealously protects its secrets. The ubiquity of caffeine in both natural and synthetic forms has been a cause of a lot of concerns among researchers and public health defenders.

Researchers have shown that caffeine increases memory [126], improves reaction time and logical reasoning, helps in periods of sleep restriction related to job and reduces drivers or pilots errors [127] and reduces risk of suicide [128] and depression [129]. It may protect against Parkinson's and Alzheimer's diseases [130]. Caffeine increases stamina during exercise [131], relieves post-workout muscle pain (cut the pain) [132] and may prevent weight gain [133]. Caffeine is beneficial in age-related chronic inflammation [134], which leads to high blood pressure, hardening of the arteries and heart diseases. It may protect against eyelid spasm [135], cataracts [136] and retinal degeneration [137], leading to blindness, against different kinds of cancer including skin cancer [138] and may prevent tinnitus (ringing in the ears) [139]. Caffeine is shown to be useful in asthma [140], lowering blood pressure [141], detoxication of the liver and the colon [142], reduction of fatty liver in non-alcoholic-related diseases [143], reduction of the liver fibrosis risk in hepatitis C [144], reduction of kidney stones risk and gout prevention [145]. It increases quality of semen in men [146] and acts as hair stimulant used in balding of men and women [147].

But due to the differences in individual sensitivity, caffeine can be easily overdosed, which may results in death—more than four cups of coffee are linked to premature death. Caffeine consumption may raise blood pressure [148], increase a risk of heart attacks among young adults [149] and gout attacks in the case of scarce caffeine overdosing [150]. It can reduce fertility [151], increase the risk of miscarriage [152], worsen the menopausal symptoms [153] and it may be a cause of breast tissue cysts in women [154]. Increased anxiety [155], depression [156], insomnia [157] and prolonged sleep deprivation problems, migraine headaches [158] are common side effects of its use. Adverse effects like incontinence [159], indigestion [160] forceful heart contractions, allergies, risk of bone fractures [161], impairment of hearing loss recovery [162], inhibition of the collagen production in the skin [163], even obesity and diabetes [164] are also on the list of potential negative effects. Recently, a large population study in the United States showed that an increase in caffeine consummation results in decrease in telomere length, which signifies accelerated ageing [165].

Many above observations, results, conclusions are mutually contradictory, which proves that despite of many years of scientific research, there are still unrevealed mysteries concerning caffeine chemical structure, physicochemical properties, its impact on living organisms, etc. Caffeine's role in producing beneficial and harmful effects is still poorly understood and definitely requires more extensive investigation.

Author details

Magdalena Latosińska and Jolanta Natalia Latosińska*

*Address all correspondence to: Jolanta.Latosinska@amu.edu.pl

Faculty of Physics, Adam Mickiewicz University, Umultowska, Poznań, Poland

References

[1] Fredholm BB, Bättig K, Holmén J, Nehlig A, Zvartau EE. Actions of caffeine in the brain with special reference to factors that contribute to its widespread use. Pharmacological Reviews. 1999;**51**(1):83-133

[2] Plants, Health and Healing. On the Interface of Ethnobotany and Medical Anthropology. Hsu E, Harris S, Editors. New York, Oxford: Berghahn Books; 2010. ISBN 978-0-85745-633-5

[3] Harbowy ME, Balentine DA, Davies AP, Cai Y. Tea chemistry. Critical Reviews in Plant Sciences. 1997;**16**(5):415-480. DOI: 10.1080/07352689709701956

[4] Heiss ML, Heiss RJ. The Story of Tea: A Cultural History and Drinking Guide. Berkeley, Calif.: Ten Speed Press; 2007. ISBN 9781580087452

[5] Lu H, Zhang J, Yang Y, Yang X, Xu B, Yang W, et al. Earliest tea as evidence for one branch of the Silk Road across the Tibetan Plateau. Scientific Reports. 2016;**6**:18955. doi: 10.1038/srep18955

[6] Forrest DM. Tea for the British: The Social and Economic History of a Famous Trade. London: Chatto & Windus; 1973. ISBN 0701119217

[7] Donaldson B. The Everything Healthy Tea Book: Discover the Healing Benefits of Tea. Avon, Massachusetts: Adams Media; 2014. ISBN 978-1440574597

[8] Epstein FT. Great Soviet Encyclopedia: A Translation of the Third Edition. Vol. 35: Association for Slavic, East European, and Eurasian Studies; 1976.ISBN 00376779

[9] Russian Tea: A Tradition Three Centuries Old. Soviet Life. 1971;**177**(6):22-23

[10] Ketchum RM. Divided Loyalties: How the American Revolution Came to New York. New York: Holt, Henry & Company, Inc; 2002. ISBN 978-0-8050-6119-2

[11] Weinberg BA, Bealer BK. The World of Caffeine: The Science and Culture of the World's Most Popular Drug. New York: Routledge; 2002. ISBN 9780203011799

[12] Caton SC. Yemen. Santa Barbara, Calif.: ABC-CLIO; 2013. ISBN 9781598849271

[13] Tibi S. Al-Razi and Islamic medicine in the 9th century. Journal of the Royal Society of Medicine. . 2006;**99**(4):206-207. DOI: 10.1258/jrsm.99.4.206

[14] Ukers WH. All about Coffee. Avon, Massachusetts: F+W Media; 2012. ISBN 9781440556326

[15] Dannenfeldt KH. Leonhard Rauwolf: Sixteenth-century Physician, Botanist, and Traveler. Bridgewater, NJ: Replica Books; 2000. ISBN 9780735102453

[16] Schwaner B, Westermann K-M. Das Wiener Kaffeehaus Legende, Kultur, Atmosphäre. Wien: Pichler; 2007. ISBN 9783854314356

[17] Doyle W. Old Regime France, 1648-1788. Oxford: Oxford Univ. Press; 2007. ISBN 978019 8731306

[18] Massie RK. Peter the Great: His Life and World. New York: Random House Inc; 2013. ISBN 9781781851289

[19] Dixon S. Catherine the Great. New York: Ecco; 2009. ISBN 9780060786274

[20] Vallee BL. Alcohol in the western world. Scientific American. 1998;**278**(6):80-85

[21] Fausto B, Fausto Sr, Brakel A. A concise history of Brazil. New York: Cambridge University Press; 2014. ISBN 9781107635241

[22] Li S., Hartland S. A new industrial process for extracting cocoa butter and xanthines with supercritical carbon dioxide. Journal of the American Oil Chemists' Society;**73**(4):423-429. doi: 10.1007/bf02523913

[23] Dillinger TL, Barriga P, Escarcega S, Jimenez M, Salazar Lowe D, Grivetti LE. Food of the gods: Cure for humanity? A cultural history of the medicinal and ritual use of chocolate. Journal of Nutrition. 2000:**130**(8S Suppl):2057s-2072s

[24] Swinnen J, Squicciarini MP. Economics of Chocolate: E-Content Generic Vendor; 2015. ISBN 9780191793264

[25] Hawkins SA. Sir Hans Sloane (1660-1735): His Life and Legacy. The Ulster Medical Journal. 2010;**79**(1):25-29

[26] Grivetti LE, Shapiro H-Y. Chocolate: History, Culture and Heritage. Hoboken, N.J.: A John Wiley & Sons; 2009. ISBN 9780470121658 0470121653

[27] The Oxford Companion to Sugar and Sweets. Oxford: Oxford University Press; 2015. ISBN 9780199313396

[28] Burdock GA, Carabin IG, Crincoli CM. Safety assessment of kola nut extract as a food ingredient. Food and Chemical Toxicology. 2009;**47**(8):1725-1732. DOI: 10.1016/j.fct.2009.04.019

[29] Bibra E, Ott J. Plant Intoxicants: A Classic Text on the Use of Mind-Altering Plants. Rochester, Vt.: Healing Arts Press; 1995. ISBN 9780892814985

[30] Pendergrast M. For God, Country, and Coca-Cola. New York: Basic Books; 2013. ISBN 9780465046997

[31] Kiple KF, Ornelas KC. The Cambridge World History of Food. Cambridge, U.K.: Cambridge University Press; 2001. ISBN 9780521402149

[32] Liu Y, Heying E, Tanumihardjo SA. History, global distribution, and nutritional importance of citrus fruits. Comprehensive Reviews in Food Science and Food Safety. 2012;**11**(6):530-545. DOI: 10.1111/j.1541-4337.2012.00201.x

[33] Stewart I. Identification of caffeine in citrus flowers and leaves. Journal of Agricultural and Food Chemistry. 1985;**33**(6):1163-1165. DOI: 10.1021/jf00066a035

[34] Yoon E, Kim J, Lee J. The U.S. consumers' acceptability and emotion measures when consuming novel Korean traditional non-alcoholic beverages. Journal of Sensory Studies. 2016;**31**(3):256-271. DOI: 10.1111/joss.12209

[35] Pybus DH, Sell CS. The chemistry of fragrances. Cambridge, UK: Royal Society of Chemistry; 1999. ISBN 9780854045280

[36] Waller GR, Macvean CD, Suzuki T. High production of caffeine and related enzyme activities in callus cultures of *Coffea arabica L*. Plant Cell Reports;**2**(3):109-112. doi: 10.1007/bf00269330

[37] Wright GA, Baker DD, Palmer MJ, Stabler D, Mustard JA, Power EF, et al. Caffeine in floral nectar enhances a pollinator's memory of reward. Science. 2013;**339**(6124):1202-1204. doi: 10.1126/science.1228806

[38] Crozier A, Ashihara H, Tomas-Barberan FA. Teas, cocoa and coffee: Plant secondary metabolites and health. Chichester, West Sussex; Hoboken, NJ: Wiley-Blackwell; 2012. ISBN 9781444347067

[39] Oudry M. Note sur la théine. Nouvelle Bibliothèque Médicale; 1827:**1**:477-479

[40] Mulder GJ. Ueber Theïn und Caffeïn. Journal für Praktische Chemie. 1838;**15**(1):280-284. DOI: 10.1002/prac.18380150124

[41] Jobst C. Thein identisch mit Caffein. Annalen der Pharmacie. 1838;**25**(1):63-66. DOI: 10.1002/jlac.18380250106

[42] Théel H. Presentation Speech by Professor Hj. Théel, President of the Swedish Royal Academy of Sciences on December 10, 1902; 1902

[43] Carlucci L, Gavezzotti A. Molecular recognition and crystal energy landscapes: An X-ray and computational study of caffeine and other methylxanthines. Chemistry. 2004;**11**(1):271-279. DOI: 10.1002/chem.200400499

[44] Bothe H, Cammenga HK. Phase transitions and thermodynamic properties of anhydrous caffeine. Journal of Thermal Analysis. 1979;**16**(2):267-275. DOI: 10.1007/BF01910688

[45] Latosińska JN, Latosińska M. Towards understanding drugs on the molecular level to design drugs with desirable profiles. In I. M. Kapetanovic (Ed.), Drug Discovery and Development - Present and Future. Rieka, Croatia: Intech; 2011. pp. 231-274. ISBN 9789533076157

[46] Descamps M, Decroix AA. Polymorphism and disorder in caffeine: Dielectric investigation of molecular mobilities. Journal of Molecular Structure. 2014;**1078**:165-173. DOI: 10.1016/j.molstruc.2014.04.042

[47] Enright GD, Terskikh VV, Brouwer DH, Ripmeester JA. The structure of two anhydrous polymorphs of caffeine from single-crystal diffraction and ultrahigh-field solid-state 13C NMR spectroscopy. Crystal Growth & Design. 2007;**7**(8):1406-1410. doi: 10.1021/cg070291o

[48] Lehmann CW, Stowasser F. The crystal structure of anhydrous beta-caffeine as determined from X-ray powder-diffraction data. Chemistry. 2007;**13**(10):2908-2911. doi: 10.1002/chem.200600973

[49] Derollez P, Correia NT, Danède F, Capet F, Affouard F, Lefebvre J, Descamps M. Ab initio structure determination of the high-temperature phase of anhydrous caffeine by X-ray powder diffraction. Acta Crystallographica B. 2005;**61**(3):329-334. doi: 10.1107/s010876810500546x

[50] Leiterer J, Emmerling F, Panne U, Christen W, Rademann K. Tracing coffee tabletop traces. Langmuir. 2008;24(15):7970-7978. DOI: 10.1021/la800768v

[51] Hedoux A, Decroix AA, Guinet Y, Paccou L, Derollez P, Descamps M. Low- and high-frequency Raman investigations on caffeine: Polymorphism, disorder and phase transformation. The Journal of Physical Chemistry B. 2011;**115**(19):5746-5753. DOI: 10.1021/jp112074w

[52] Larkin PJ, Dabros M, Sarsfield B, Chan E, Carriere JT, Smith BC. Polymorph characterization of active pharmaceutical ingredients (APIs) using low-frequency Raman spectroscopy. Applied Spectroscopy. 2014;**68**(7):758-776. DOI: 10.1366/13-07329

[53] Seliger J, Žagar V, Apih T, Gregorovic A, Latosińska M, Olejniczak GA, Latosińska JN. Polymorphism and disorder in natural active ingredients. Low and high-temperature phases of anhydrous caffeine: Spectroscopic ((1)H-(14)N NMR-NQR/(14)N NQR) and solid-state computational modelling (DFT/QTAIM/RDS) study. European Journal of Pharmaceutical Sciences. 2016;**85**:18-30. DOI: 10.1016/j.ejps.2016.01.025

[54] Latosińska JN, Latosińska M, Olejniczak GA, Seliger J, Žagar V. Topology of the interactions pattern in pharmaceutically relevant polymorphs of methylxanthines (caffeine, theobromine, and theophiline): Combined experimental ((1)H-(1)(4)N nuclear quadrupole double resonance) and computational (DFT and Hirshfeld-based) study. Journal of Chemical Information and Modeling. 2014;**54**(9):2570-2584. DOI: 10.1021/ci5004224

[55] Fulgoni VL 3rd, Keast DR, Lieberman HR. Trends in intake and sources of caffeine in the diets of US adults: 2001-2010. The American Journal of Clinical Nutrition. 2015;**101**(5):1081-1087. DOI: 10.3945/ajcn.113.080077

[56] Grobbee DE, Rimm EB, Giovannucci E, Colditz G, Stampfer M, Willett W. Coffee, caffeine, and cardiovascular disease in men. The New England Journal of Medicine. 1990;**323**(15):1026-1032. doi: 10.1056/nejm199010113231504

[57] Jee SH, He, J, Whelton PK, Suh I, Klag MJ. The effect of chronic coffee drinking on blood pressure: A meta-analysis of controlled clinical trials. Hypertension. 1999;**33**(2):647-652. doi: 10.1161/01.hyp.33.2.647

[58] Robertson D, Hollister AS, Kincaid D, Workman R, Goldberg MR, Tung CS, Smith B. Caffeine and hypertension. The American Journal of Medicine. 1984;**77**(1):54-60. doi: 10.1016/0002-9343(84)90435-2

[59] Nurminen M-L, Niittynen L, Korpela R, Vapaatalo H. Coffee, caffeine and blood pressure: a critical review. European Journal of Clinical Nutrition. 1999;**53**(11):831-839. doi: 10.1038/sj.ejcn.1600899

[60] Arnaud MJ. The pharmacology of caffeine. Progress in Drug Research. 1987;**31**:273-313

[61] Berthou F, Guillois B, Riche C, Dreano Y, Jacqz-Aigrain E, Beaune PH. Interspecies variations in caffeine metabolism related to cytochrome P4501A enzymes. Xenobiotica. 1992;**22**(6):671-680. DOI: 10.3109/00498259209053129

[62] Ashihara H, Sano H, Crozier A. Caffeine and related purine alkaloids: biosynthesis, catabolism, function and genetic engineering. Phytochemistry. 2008;**69**(4):841-856. DOI: 10.1016/j.phytochem.2007.10.029

[63] Caffeine. (1991/01/01 ed. Vol. 51). Lyon: IARC; 1991. ISBN 1017-1606.

[64] Tassaneeyakul W, Birkett DJ, McManus ME, Tassaneeyakul W, Veronese ME, Andersson T, Miners JO. Caffeine metabolism by human hepatic cytochromes P450: contributions of 1A2, 2E1 and 3A isoforms. Biochemical Pharmacology. 1994;**47**(10):1767-1776. DOI: 10.1016/0006-2952(94)90304-2

[65] Smith BD, Gupta U, Gupta BS. Caffeine and activation theory: Effects on health and behavior. Boca Raton: CRC Press; 2007.ISBN 0-8493-7102-3

[66] Orru M, Guitart X, Karcz-Kubicha M, Solinas M, Justinova Z, Barodia SK, et al. Psychostimulant pharmacological profile of paraxanthine, the main metabolite of caffeine in humans. Neuropharmacology. 2013;**67**:476-484. DOI: 10.1016/j.neuropharm.2012.11.029

[67] Smit HJ. Theobromine and the pharmacology of cocoa. The Handbook of Experimental Pharmacology(200); 2011:201-234. DOI: 10.1007/978-3-642-13443-2_7

[68] Dubuis E, Wortley MA, Grace MS, Maher SA, Adcock JJ, Birrell MA, Belvisi MG. Theophylline inhibits the cough reflex through a novel mechanism of action. Journal of Allergy and Clinical Immunology. 2014;**133**(6):1588-1598. DOI: 10.1016/j.jaci.2013.11.017

[69] Petzer JP, Castagnoli N, Schwarzschild MA, Chen J-F, Van der Schyf CJ. Dual-target-directed drugs that block monoamine oxidase B and adenosine A(2A) receptors for Parkinson's disease. Neurotherapeutics. 2009;**6**(1):141-151. doi: 10.1016/j.nurt.2008.10.035

[70] Lopez F, Miller LG, Greenblatt DJ, Kaplan GB, Shader RI. Interaction of caffeine with the GABAA receptor complex: alterations in receptor function but not ligand binding. The European Journal of Pharmacology. 1989;**172**(6):453-459. DOI: 10.1016/0922-4106(89)90028-X

[71] Ferre S, Orru M, Guitart X. Paraxanthine: Connecting caffeine to nitric oxide neurotransmission. Journal of Caffeine Research. 2013;**3**(2):72-78. DOI: 10.1089/jcr.2013.0006

[72] Ross GW, Abbott RD, Petrovitch H, Morens DM, Grandinetti A, Tung KH, et al. Association of coffee and caffeine intake with the risk of Parkinson disease. Journal of the American Medical Association. 2000;**283**(20):2674-2679. DOI: 10.1001/jama.283.20.2674

[73] Postuma RB, Lang AE, Munhoz RP, Charland K, Pelletier A, Moscovich M, et al. Caffeine for treatment of Parkinson disease: A randomized controlled trial. Neurology. 2012;**79**(7):651-658. doi: 10.1212/WNL.0b013e318263570d

[74] Popat RA, Van Den Eeden SK, Tanner CM, Kamel F, Umbach DM, Marder K, et al. Coffee, ADORA2A, and CYP1A2: the caffeine connection in Parkinson's disease. European Journal of Neurology. 2011;**18**(5):756-765. doi: 10.1111/j.1468-1331.2011.03353.x

[75] Eskelinen MH, Kivipelto M. Caffeine as a protective factor in dementia and Alzheimer's disease. Journal of Alzheimer's Disease. 2010;**20**:167-174. DOI: 10.3233/jad-2010-1404

[76] Maia L, Mendonca A. Does caffeine intake protect from Alzheimer's disease? European Journal of Neurology. 2002;**9**(4):377-382. doi: 10.1046/j.1468-1331.2002.00421.x

[77] Biessels GJ. Caffeine, diabetes, cognition, and dementia. Journal of Alzheimer's Disease. 2010;**20**:143-150. DOI: 10.3233/jad-2010-091228

[78] Chen J-F, Chern Y. Impacts of methylxanthines and adenosine receptors on neurodegeneration: Human and experimental studies. The Handbook of Experimental Pharmacology(200). 2011:267-310. doi: 10.1007/978-3-642-13443-2_10

[79] Guerreiro S, Toulorge D, Hirsch E, Marien M, Sokoloff P, Michel PP. Paraxanthine, the primary metabolite of caffeine, provides protection against dopaminergic cell death via stimulation of ryanodine receptor channels. Molecular Pharmacology. 2008;**74**(4):980-989. DOI: 10.1124/mol.108.048207

[80] Costentin J. Main neurotropic and psychotropic effects of methylxanthines (caffeine, theophylline, theobromine, paraxanthine). Psychiatrie Sciences Humaines Neurosciences. 2010;**8**(4):182-186. DOI: 10.1007/s11836-010-0141-z

[81] Dórea JG, da Costa THM. Is coffee a functional food? British Journal of Nutrition. 2005;**93**(06):773-782. DOI: 10.1079/bjn20051370

[82] Nechay BR. Potentiation of diuretic effects of methyl xanthines and pyrimidines by carbonic anhydrase inhibitors. Journal of Pharmacology and Experimental Therapeutics. 1964;**144**:276-283

[83] Dorfman LJ, Jarvik ME. Comparative stimulant and diuretic actions of caffeine and theobromine in man. Clinical Pharmacology & Therapeutics. 1970;**11**:869-872. DOI: 10.1002/cpt1970116869

[84] Osei KA, Ovenseri-Ogbomo G, Kyei S, Ntodie M. The effect of caffeine on tear secretion. Optometry and Vision Science. 2014;**91**:171-177. DOI: 10.1097/OPX.0000000000000129

[85] Moss SE, Klein R, Klein BE. Prevalence of and risk factors for dry eye syndrome. Archives of Ophthalmology. 2000;**118**:1264-1268. DOI: 10.1001/archopht.118.9.1264

[86] Nkondjock A. Coffee consumption and the risk of cancer: An overview. Cancer Letters. 2009;**277**(2):121-125. doi: 10.1016/j.canlet.2008.08.022

[87] Bøhn SK, Blomhoff R, Paur I. Coffee and cancer risk, epidemiological evidence, and molecular mechanisms. Molecular Nutrition & Food Research. 2014;**58**(5):915-930. doi: 10.1002/mnfr.201300526

[88] Song F, Qureshi AA, Han J. Increased caffeine intake is associated with reduced risk of basal cell carcinoma of the skin. Cancer Research. 2012;**72**(13):3282-3289. doi: 10.1158/0008-5472.can-11-3511

[89] Li G, Ma D, Zhang Y, Zheng W, Wang P. Coffee consumption and risk of colorectal cancer: a meta-analysis of observational studies. Public Health Nutrition. 2013;**16**(2):346-357. doi: 10.1017/s1368980012002601

[90] Hashibe M, Galeone C, Buys SS, Gren L, Boffetta P, Zhang ZF, La Vecchia C. Coffee, tea, caffeine intake, and the risk of cancer in the PLCO cohort. British Journal of Cancer. 2015. doi: 10.1038/bjc.2015.276

[91] Bertrand B, Stefan L, Pirrotta M, Monchaud D, Bodio E, Richard P, et al. Caffeine-based Gold(I) N-heterocyclic carbenes as possible anticancer agents: Synthesis and biological properties. Inorganic Chemistry. 2014;**53**(4):2296-2303. DOI: 10.1021/ic403011h

[92] Kronschläger M, Löfgren S, Yu Z, Talebizadeh N, Varma SD, Söderberg P. Caffeine eye drops protect against UV-B cataract. Experimental Eye Research. 2013;**113**(0):26-31. DOI: 10.1016/j.exer.2013.04.015

[93] Heffernan TP, Kawasumi M, Blasina A, Anderes K, Conney AH, Nghiem P. ATR-Chk1 pathway inhibition promotes apoptosis after UV treatment in primary human keratinocytes: Potential basis for the UV protective effects of caffeine. The Journal of Investigative Dermatology. 2009;**129**(7):1805-1815. doi: 10.1038/jid.2008.435

[94] Kerzendorfer C, O'Driscoll M. UVB and caffeine: inhibiting the DNA damage response to protect against the adverse effects of UVB. The Journal of Investigative Dermatology. 2009;**129**(7):1611-1613. DOI: 10.1038/jid.2009.99

[95] Lu Y-P, Lou Y-R, Xie J-G, Peng Q-Y, Liao J, Yang CS, et al. Topical applications of caffeine or (-)-epigallocatechin gallate (EGCG) inhibit carcinogenesis and selectively increase apoptosis in UVB-induced skin tumors in mice. Proceedings of the National Academy of Sciences of the United States of America. 2002;**99**(19):12455-12460. doi: 10.1073/pnas.182429899

[96] Sabisz M, Skladanowski A. Modulation of cellular response to anticancer treatment by caffeine: Inhibition of cell cycle checkpoints, DNA repair and more. Current Pharmaceutical Biotechnology;**9**(4):325-336. doi: 10.2174/138920108785161497

[97] Ohta A, Sitkovsky M. The adenosinergic immuno-modulatory drugs. Current Opinion in Pharmacology. 2009;**9**(4):501-506. DOI: 10.1016/j.coph.2009.05.005

[98] Zigman S, Schultz J, Yulo T. Possible roles of near UV light in the cataractous process. Experimental Eye Research. 1973;15(2):201-208. DOI: 10.1016/0014-4835(73)90120-6

[99] Conterio F, Chiarelli B. Study of the inheritance of some daily life habits. Heredity (Edinburgh). 1962;**17**:347-359. DOI: 10.1038/hdy.1962.36

[100] Kendler KS, Prescott CA. Caffeine intake, tolerance, and withdrawal in women: A population-based twin study. The American Journal of Psychiatry. 1999;**156**(2):223-228. DOI: 10.1176/ajp.156.2.223

[101] Goodman MT, Tung KH, McDuffie K, Wilkens LR, Donlon TA. Association of caffeine intake and CYP1A2 genotype with ovarian cancer. Nutrition and Cancer. 2003;**46**(1):23-29. DOI: 10.1207/s15327914nc4601_03

[102] Kotsopoulos J, Ghadirian P, El-Sohemy A, Lynch HT, Snyder C, Daly M, et al. The CYP1A2 genotype modifies the association between coffee consumption and breast cancer risk among BRCA1 mutation carriers. Cancer Epidemiology, Biomarkers & Prevention. 2007;**16**(5):912-916. DOI: 10.1158/1055-9965.epi-06-1074

[103] Sachse C, Brockmoller J, Bauer S, Roots I. Functional significance of a C-->A polymorphism in intron 1 of the cytochrome P450 CYP1A2 gene tested with caffeine. British Journal of Clinical Pharmacology. 1999;**47**(4):445-449. DOI: 10.1046/j.1365-2125.1999.00898.x

[104] Yang A, Palmer AA, de Wit H. Genetics of caffeine consumption and responses to caffeine. Psychopharmacology (Berlin). 2010;**211**(3):245-257. DOI: 10.1007/s00213-010-1900-1

[105] Cornelis MC, Monda KL, Yu K, Paynter N, Azzato EM, Bennett SN, et al. Genome-wide meta-analysis identifies regions on 7p21 (AHR) and 15q24 (CYP1A2) as determinants of habitual caffeine consumption. PLOS Genetics. 2011;**7**(4):e1002033. doi: 10.1371/journal.pgen.1002033

[106] Josse AR, Da Costa LA, Campos H, El-Sohemy A. Associations between polymorphisms in the AHR and CYP1A1-CYP1A2 gene regions and habitual caffeine consumption. The American Journal of Clinical Nutrition. 2012;**96**(3):665-671. DOI: 10.3945/ajcn.112.038794

[107] Retey JV, Adam M, Khatami R, Luhmann UF, Jung HH, Berger W, Landolt HP. A genetic variation in the adenosine A2A receptor gene (ADORA2A) contributes to individual sensitivity to caffeine effects on sleep. Clinical Pharmacology & Therapeutics. 2007;**81**(5):692-698. DOI: 10.1038/sj.clpt.6100102

[108] Happonen P, Voutilainen S, Tuomainen T-P, Salonen JT. Catechol-O-methyltransferase gene polymorphism modifies the effect of coffee intake on incidence of acute coronary events. PLOS ONE. 2006;**1**(1):e117. doi: 10.1371/journal.pone.0000117

[109] Kohlmeier M. Nutrigenetics: Applying the Science of Personal Nutrition: Academic Press; 2012. ISBN 9780123859013

[110] Cornelis MC, El-Sohemy A, Campos H. Genetic polymorphism of the adenosine A2A receptor is associated with habitual caffeine consumption. The American Journal of Clinical Nutrition. 2007;**86**(1):240-244

[111] Peters JM. Factors affecting caffeine toxicity: A review of the literature. The Journal of Clinical Pharmacology and The Journal of New Drugs. 1967;**7**(3):131-141. DOI: 10.1002/j.1552-4604.1967.tb00034.x

[112] Nawrot P, Jordan S, Eastwood J, Rotstein J, Hugenholtz A, Feeley M. Effects of caffeine on human health. Food Additives & Contaminants. 2003;**20**(1):1-30. DOI: 10.1080/0265203021000007840

[113] Stavric B. Methylxanthines: Toxicity to humans. 2. Caffeine. Food and Chemical Toxicology. 1988;**26**(7):645-662. DOI: 10.1016/0278-6915(88)90236-0

[114] James JE. Caffeine and health. London: Academic Press; 1991. ISBN 9780123801050

[115] Stidworthy MF, Bleakley JS, Cheeseman MT, Kelly DF. Chocolate poisoning in dogs. Veterinary Record. 1997;**141**(1):28

[116] Sutton RH. Cocoa poisoning in a dog. Veterinary Record. 1981;**109**(25-26):563-564

[117] Kovalkovičová N, Šutiaková I, Pistl J, Šutiak V. Some food toxic for pets. Interdisciplinary Toxicology. 2009;**2**(3):169-176. DOI: 10.2478/v10102-009-0012-4

[118] Watson RR, Preedy VR, Zibadi S. Chocolate in health and nutrition: Springer; 2013. ISBN 9781617798030

[119] Gartrell BD, Reid C. Death by chocolate: A fatal problem for an inquisitive wild parrot. The New Zealand Veterinary Journal. 2007;**55**(3):149-151. DOI: 10.1080/00480169.2007.36759

[120] Lightfoot TL, Yeager JM. Pet bird toxicity and related environmental concerns. Veterinary Clinics of North America: Exotic Animal Practice. 2008;**11**(2):229-259. DOI: 10.1016/j.cvex.2008.01.006

[121] Brandon DM, Bicket TL. Hugs for cat lovers. New York: Howard Books; 2014.ISBN 9781416560340

[122] Hollingsworth RG, Armstrong JW, Campbell E. Caffeine as a repellent for slugs and snails. Nature. 2002;**417**(6892):915-916. DOI: 10.1038/417915a

[123] Nathanson JA. Caffeine and related methylxanthines: possible naturally occurring pesticides. Science. 1984;**226**(4671):184-187. DOI: 10.1126/science.6207592

[124] Noever DA, Cronise RJ, Relwani RA. Using spider-web patterns to determine toxicity. NASA Technical Briefs. 1995;**19**(4):82

[125] Jaramillo J, Borgemeister C, Baker P. Coffee berry borer Hypothenemus hampei (Coleoptera: Curculionidae): searching for sustainable control strategies. Bulletin of Entomological Research. 2006;**96**(3):223-233. DOI: 10.1079/BER2006434

[126] Borota D, Murray E, Keceli G, Chang A, Watabe JM, Ly M, et al. Post-study caffeine administration enhances memory consolidation in humans. Nature Neuroscience. 2014;**17**(2):201-203. DOI: 10.1038/nn.3623

[127] Horne JA, Reyner LA. Counteracting driver sleepiness: effects of napping, caffeine, and placebo. Psychophysiology. 1996;**33**(3):306-309. DOI: 10.1111/j.1469-8986.1996.tb00428.x

[128] Lucas M, O'Reilly EJ, Pan A, Mirzaei F, Willett WC, Okereke OI, Ascherio A. Coffee, caffeine, and risk of completed suicide: Results from three prospective cohorts of American adults. The World Journal of Biological Psychiatry. 2014;**15**(5):377-386. DOI: 10.3109/15622975.2013.795243

[129] Lucas M, Mirzaei F, Pan A, et al. Coffee, caffeine, and risk of depression among women. Archives of Internal Medicine. 2011;**171**(17):1571-1578. DOI: 10.1001/archinternmed.2011.393

[130] Cao C, Cirrito JR, Lin X, Wang L, Verges DK, Dickson A, et al. Caffeine suppresses β-amyloid levels in plasma and brain of Alzheimer's transgenic mice. Journal of Alzheimer's Disease. 2009;**17**(3):681-697. DOI: 10.3233/JAD-2009-1071

[131] Tauler P, Martínez S, Moreno C, Monjo M, Martínez P, Aguiló A. Effects of caffeine on the inflammatory response induced by a 15-km run competition. Medicine and Science in Sports and Exercise. 2013;**45**(7):1269-1276. DOI: 10.1249/mss.0b013e3182857c8a

[132] Maridakis V, O'Connor PJ, Dudley GA, McCully KK. Caffeine attenuates delayed-onset muscle pain and force loss following eccentric exercise. The Journal of Pain;**8**(3):237-243. DOI: 10.1016/j.jpain.2006.08.006

[133] Icken D, Feller S, Engeli S, Mayr A, Muller A, Hilbert A, de Zwaan M. Caffeine intake is related to successful weight loss maintenance. European Journal of Clinical Nutrition. 2016;**70**(4):532-534. DOI: 10.1038/ejcn.2015.183

[134] Furman D, Chang J, Lartigue L, Bolen CR, Haddad F, Gaudilliere B, Faustin B. Expression of specific inflammasome gene modules stratifies older individuals into two extreme clinical and immunological states. Nature Medicine. 2017;**23**(2):174-184. DOI: 10.1038/nm.4267

[135] Defazio G, Martino D, Abbruzzese G, Girlanda P, Tinazzi M, Fabbrini G, Berardelli A. Influence of coffee drinking and cigarette smoking on the risk of primary late onset blepharospasm: Evidence from a multicentre case control study. Journal of Neurology, Neurosurgery, and Psychiatry. 2007;**78**(8):877-879. DOI: 10.1136/jnnp.2007.119891

[136] Varma SD, Kovtun S, Hegde K. Effectiveness of topical caffeine in cataract prevention: Studies with galactose cataract. Molecular Vision. 2010;**16**:2626-2633

[137] Jang H, Ahn HR, Jo H, Kim K-A, Lee EH, Lee KW, et al. Chlorogenic acid and coffee prevent hypoxia-induced retinal degeneration. Journal of Agricultural and Food Chemistry. 2014;**62**(1):182-191. DOI: 10.1021/jf404285v

[138] Loftfield E, Mayne S, Shebl F, Freedman N, Graubard B, Sinha R. Abstract LB-280: Prospective study of coffee drinking and risk of melanoma in the United States. Cancer Research. 2014;**74**(19):LB-280-LB-280. doi: 10.1158/1538-7445.am2014-lb-280

[139] Glicksman JT, Curhan SG, Curhan GC. A prospective study of caffeine intake and risk of incident tinnitus. The American Journal of Medicine;**127**(8):739-743. DOI: 10.1016/j.amjmed.2014.02.033

[140] Welsh EJ, Bara A, Barley E, Cates, CJ. Caffeine for asthma. Cochrane Database of Systematic Reviews. 2010;(1). doi: 10.1002/14651858.CD001112.pub2

[141] Hodgson JM, Puddey IB, Burke V, Beilin LJ, Jordan N. Effects on blood pressure of drinking green and black tea. Journal of Hypertension. 1999;**17**(4):457-463. DOI: 10.1097/00004872-199917040-00002

[142] Teekachunhatean S, Tosri N, Rojanasthien N, Srichairatanakool S, Sangdee C. Pharmacokinetics of caffeine following a single administration of coffee enema versus oral coffee consumption in healthy male subjects. ISRN Pharmacology. 2013;**2013**:147238. doi: 10.1155/2013/147238

[143] Sinha RA, Farah BL, Singh BK, Siddique MM, Li Y, Wu Y, et al. Caffeine stimulates hepatic lipid metabolism by the autophagy-lysosomal pathway in mice. Hepatology. 2014;**59**(4):1366-1380. DOI: 10.1002/hep.26667

[144] Khalaf N, White D, Kanwal F, Ramsey D, Mittal S, Tavakoli-Tabasi S,et al. Coffee and caffeine are associated with decreased risk of advanced hepatic fibrosis among patients with hepatitis C. Clinical Gastroenterology and Hepatology. 2015;**13**(8):1521-1531.e1523. doi: 10.1016/j.cgh.2015.01.030

[145] Choi HK, Willett W, Curhan G. Coffee consumption and risk of incident gout in men: A prospective study. Arthritis & Rheumatism. 2007;**56**(6):2049-2055. DOI: 10.1002/art.22712

[146] Karmon AE, Toth TL, Chiu YH, Gaskins AJ, Tanrikut C, Wright DL. Male caffeine and alcohol intake in relation to semen parameters and in vitro fertilization outcomes among fertility patients. Andrology. 2017;**5**(2):354-361. DOI: 10.1111/andr.12310

[147] Fischer TW, Hipler UC, Elsner P. Effect of caffeine and testosterone on the proliferation of human hair follicles in vitro. International Journal of Dermatology. 2007;**46**(1):27-35. DOI: 10.1111/j.1365-4632.2007.03119.x

[148] Vlachopoulos C, Hirata K, Stefanadis C, Toutouzas P, O'Rourke MF. Caffeine increases aortic stiffness in hypertensive patients. American Journal of Hypertension. 2003;**16**(1):63-66. DOI: 10.1016/S0895-7061(02)03155-2

[149] Mos L, Fania C, Benetti E, Bratti P, Maraglino G, Mazzer A, et al. Coffee consumption is a predictor of cardiovascular events in young and middle aged hypertensive subjects. Journal of Hypertension. 2015;**33**:10. doi: 10.1097/01.hjh.0000467378.82732.55

[150] Choi HK, Curhan G. Coffee consumption and risk of incident gout in women: The Nurses' Health Study. The American Journal of Clinical Nutrition. 2010;**92**(4):922-927. DOI: 10.3945/ajcn.2010.29565

[151] Cao H, Ren J, Feng X, Yang G, Liu J. Is caffeine intake a risk factor leading to infertility? A protocol of an epidemiological systematic review of controlled clinical studies. Systematic Reviews. 2016;**5**:45. doi: 10.1186/s13643-016-0221-9

[152] Hahn KA, Wise LA, Rothman KJ, Mikkelsen EM, Brogly SB, Sørensen HT, Hatch EE. Caffeine and caffeinated beverage consumption and risk of spontaneous abortion. Human Reproduction. 2015;**30**(5):1246-1255. DOI: 10.1093/humrep/dev063

[153] Faubion SS, Sood R, Thielen JM, Shuster LT. Caffeine and menopausal symptoms: what is the association? Menopause. 2015;**22**(2):155-158. DOI: 10.1097/gme.0000000000000301

[154] Boyle CA, Berkowitz GS, LiVolsi VA, Ort S, Merino MJ, White C, Kelsey JL. Caffeine consumption and fibrocystic breast disease: A case-control epidemiologic study. Journal of the National Cancer Institute. 1984;**72**(5):1015-1019. DOI: 10.1093/jnci/72.5.1015

[155] Veleber DM, Templer DI. Effects of caffeine on anxiety and depression. The Journal of Abnormal Psychology. 1984;**93**(1):120-122. DOI: 10.1037/0021-843X.93.1.120

[156] Richards G, Smith AP. A review of energy drinks and mental health, with a focus on stress, anxiety, and depression. Journal of Caffeine Research. 2016;**6**(2):49-63. DOI: 10.1089/jcr.2015.0033

[157] Karacan I, Thornby JI, Anch M, Booth GH, Williams RL, Salis PJ. Dose-related sleep disturbances induced by coffee and caffeine. Clinical Pharmacology and Therapeutics. 1976;**20**(6):682-689. DOI: 10.1002/cpt1976206682

[158] Scher AI, Stewart WF, Lipton RB. Caffeine as a risk factor for chronic daily headache: A population-based study. Neurology. 2004;**63**(11):2022-2027. DOI: 10.1046/j.1365-2125.1999.00898.x

[159] Gleason JL, Richter HE, Redden DT, Goode PS, Burgio KL, Markland AD. Caffeine and urinary incontinence in US women. International Urogynecology Journal. 2013;**24**(2):295-302. DOI: 10.1007/s00192-012-1829-5

[160] Shirlow MJ, Mathers CD. A study of caffeine consumption and symptoms; indigestion, palpitations, tremor, headache and insomnia. International Journal of Epidemiology. 1985;**14**(2):239-248. DOI: 10.1093/ije/14.2.239

[161] Tomaszewski M, Olchowik G, Tomaszewska M, Dworzanski W, Burdan F. The influence of caffeine administered at 10 degrees C on bone tissue development. The Annals of Agricultural and Environmental Medicine. 2016;**23**(2):319-323. DOI: 10.5604/12321966.1203898

[162] Zawawi F, Bezdjian A, Mujica-Mota M, Rappaport J, Daniel SJ. Association of caffeine and hearing recovery after acoustic overstimulation events in a guinea pig model. Otolaryngology–Head & Neck Surgery. 2016;**142**(4):383-388. DOI: 10.1001/jamaoto.2015.3938

[163] Donejko M, Przylipiak A, Rysiak E, Głuszuk K, Surażyński A. Influence of caffeine and hyaluronic acid on collagen biosynthesis in human skin fibroblasts. Drug Design, Development and Therapy. 2014;**8**:1923-1928. DOI: 10.2147/DDDT.S69791

[164] Malik VS, Popkin BM, Bray GA, Després J-P, Willett WC, Hu FB. Sugar-sweetened beverages and risk of metabolic syndrome and type 2 diabetes: A meta-analysis. Diabetes Care. 2010;**33**(11):2477-2483. DOI: 10.2337/dc10-1079

[165] Tucker LA. Caffeine consumption and telomere length in men and women of the National Health and Nutrition Examination Survey (NHANES). Nutrition & Metabolism. 2017;**14**(1):10. doi: 10.1186/s12986-017-0162-x

Chapter 2

Coffea arabica: A Plant with Rich Content in Caffeine

Eva Brigitta Patay, Luminița Fritea,
Andreea Antonescu, Angela Antonescu and
Luciana Dobjanschi

Additional information is available at the end of the chapter

http://dx.doi.org/10.5772/intechopen.68149

Abstract

Coffea arabica L. is the most well-known and studied *Coffea* taxa, which is very popular in both scientific and social fields. This comprehensive work was created in order to describe its phytochemical composition and to present the metabolism of caffeine, which is the most important alkaloid from this plant. The analytical methods used for caffeine determination such as chromatographic, electrochemical, and spectroscopic techniques are also presented. In addition, this work emphasizes the medicinal importance of caffeine, which can present both important beneficial and secondary effects for human body.

Keywords: caffeine, coffee, phytochemistry, analytical methods, *Coffea arabica*, medicinal importance

1. Introduction

Coffea species are well-known tropical plants, which are mostly used for preparing the famous beverage called coffee. Coffee plants present a tremendous importance in the scientific, agricultural, social, and commercial fields being on the second place after petrol in the international market [1].

The genus *Coffea* belongs to the Rubiaceae family and comprises up to 124 species. The most famous and used species are *Coffea arabica* L., *Coffea robusta* L. Linden (syn.: *Coffea canephora* Pierre ex A. Froehner), and *Coffea liberica* Hiern. (syn.: *Coffea dewevrei* De Wild. & T. Durand) [2, 3]. Nowadays, the Arabic coffee makes up to 75% from the total coffee production of the world; meanwhile, the Robusta coffee makes up to 24%, and the Liberian coffee 1%, respectively [4].

There are some legends about the discovery of coffee seeds and their utilization as beverage and food. The coffee seeds initially were consumed as snacks. The seeds were mixed with animal fat, and these were eaten by people during their long trips [5]. A kind of wine was also obtained from the fermented fruits of coffee [6]. One story tells that in the fifteenth century, there was a goat herder who noticed the stimulating effect of the coffee fruits when his goats ate these berries. He went to the nearest monastery to tell the monks his experience and they started to drink it during their long morning prayers [5].

The caffeine consumption has a long history dating back to antiquity. There are more than 60 plant species, which have been identified as containing caffeine. Plants like coffee, guarana, yoco, mate, and cassina were successfully used for the preparation of caffeine-containing beverages. According to history, tea is the oldest caffeine-containing beverage which had already been mentioned by the Chinese emperor Shen Nung in 2737 B.C., and it had been listed in a Chinese dictionary in about 350 AD [6].

The Arabic countries used to prohibit the consumption of coffee among men by referring to the Koran. In the sixteenth century, special places were established in Constantinople where Turkish dignitaries were served with coffee. As Turkish men spent a lot of time in these establishments, the mosques began to be abandoned, so imams opposed to the coffee consumption, which was forbidden by a law ordered by Muhammad. Due to this reason, Sultan Murad banned the coffee consumption, but later this law was modified when the use of tobacco started to become more popular [7].

Nowadays, approximately 7 million tons of coffee is used for the preparation of the famous coffee beverage according to Food and Agriculture Organization of the United Nations [8].

2. *Coffea arabica* L.: geographical distribution, cultivation, and processing

Coffea species belong to the Rubiaceae family, which includes 450 genera and more than 6500 species. *Coffea* species are originated from Africa and Madagascar being cultivated in both Ethiopian and Yemen regions as the geographical and climate conditions are the same. Then, they have also spread in other continents and nowadays they are present in all the tropical and subtropical regions, including Australia and China. It is interesting that wild species do not occur in America, native taxa are found only in Africa and in the South Asian regions [2, 9]. They grow in forests at an altitude of 950–2000 m and at 18–22°C, the most favorable altitude being between 1300 and 1600 m [2, 3, 10]. There are also coffee shrubs, which are cultivated at higher regions, and they are named "high grown" coffee with an outstanding quality [11].

The coffee crops are spread in Central and South America, Africa, and Asia leading to a total of about 11 million hectares of plantations and an annual yield of about 6 million tons of green coffee [11]. *C. arabica* is often cultivated in large-scale plantations and in small holder farms too. In addition, they are also planted on cooler mountain ranges, at the base of shady trees, which protect them against strong sunlight. A shrub plant yields well for more than 20 years, and the ripe and red fruits are individually hand-picked [12].

Coffee products are processed in two methods: wet and dry processes. The dry procedure is popular in Brazil and the other one in Central America. During the dry method, the fruits are dried in open air for 2–3 weeks before the fermentation step [11, 13]. In addition, the wet method can be used only for ripe fruit as the fermentation positively affects the quality of seeds, external morphological aspects, and the taste [11, 13].

The roasting process, which is responsible for the characteristic aroma of the seeds, takes place only in customer country at 200–250°C [14]. This roasting process presents three phases such as an initial drying phase (during this endothermic step, the wet is eliminated and the color is turned yellowish), the roasting phase (based on several complex pyrolytic reactions resulting in many chemical compounds which will confer the coffee aroma and taste, the beans are transformed to dark brown during exothermic and endothermic phases), and the cooling phase (using air or water in order to end the last exothermic process [14].

3. Phytochemical features of *Coffea arabica* L.

The plant part of the coffee being officially used is the seed (*Coffeae semen*), which contains among other compounds polyphenols and alkaloids as biologically active substances. These two groups of chemicals made coffee popular in both scientific and social fields. The chemical composition of coffee seeds contains a large variety of substances among which the most representative groups are purine alkaloids (caffeine, theobromine, theophylline), polyphenols (chlorogenic acid, caffeic acid, etc.), alkaloids (trigonelline), fatty oil, carotenoids, enzymes, phytosterols (sitosterol, dihydro-sitosterol, stigmasterol), tannin, wax, carbohydrates as monosaccharides (fructose, glucose, etc.), oligosaccharides (sucrose, etc.) and polymers (cellulose, hemicelluloses), and nonvolatile and volatile aliphatic acids (citric, oxalic, acetic, isovaleric, decanoic acids, etc.) [2, 15]. In addition, they contain also some minerals such as K (40%), P (4%), Na, Mg (with variation between species), Ca, and S; and trace minerals with a variation according to the soil composition such as Zn, Sr, Mn, Fe, Cu, Ba, B, and Al. Nicotinic acid (vitamin PP) is formed from trigonelline demethylation during roasting process [15].

The characteristic aroma of coffee appears during the roasting process. The principal volatile compounds include derivatives of: sulfur (thiols, hydrogen sulfide, thiophenes, thiazoles), pyrazine (pyrazine itself, thiol and furfuryl derivatives, alkyl derivatives), pyridine, pyrrole, oxazole, furan, aldehydes, ketones, and phenols [14, 15]. Caffeine is present in all plant parts with the highest concentration in the immature seeds. This alkaloid with a bitter taste can be quantitatively reduced during the roasting and decaffeination processes [16].

Caffeine can present autotoxic and inhibiting effect on the mitosis and cell plate formation in rootlet. Studies demonstrated that the cell divisions in root tips started only after that it was pushed away from the caffeine-rich endosperm by elongation of the hypocotyl and maintained through the cell elongation. Caffeine is introduced into the embryonic cotyledons mostly after the cell division is completed there [17]. Some authors suggested that the physiological significance of caffeine could be the prevention of predation by animals and these alkaloids function as allochemicals in the pericarp and seeds [18]. Josef et al. mentioned in their work about insecticidal effect of caffeine, which can also synergize the effects of pesticides [19].

Polyphenols are another representative group from the coffee seeds composition containing kaempferol, quercetin, ferulic, sinapic, nicotinic, quinolic, tannic, and pyrogallic acids which present important effects such as antioxidant, hepatoprotective, antibacterial, antiviral, anti-inflammatory, and hypolipidemic ones [2, 20–27]. In addition, caffeic, chlorogenic, p-coumaric, ferulic and sinapic acids, rutin, quercetin, kaempferol, and isoquercitrin were identified in fruits of *C. arabica* and *C. benghalensis* Roxb. ex Schult. The nonhydrolyzed extract of the pericarp of both species presented important quantities of chlorogenic acids [28].

In one of our previous study, a high phenolic content was observed in the immature pericarp of *C. benghalensis* and *C. liberica* compared with that of *C. arabica.* In addition, the immature pericarp of Bengal coffee and the immature seed of Arabic coffee showed a significant polyphenol content too [29]. Other phytochemical studies showed that the total hydroxycinnamic acid content of *C. arabica* was significantly higher than that of *Coffea sessiliflora* Bridson, *Coffea resinora* Hook.f., and *Coffea leroyi* A.P.Davis. The mangiferin, isomangiferin and caffeoylquinic acid were present in higher concentration in the young leaves than in other plant parts [30–32]. *Coffea anthonyi* Stoff. & F. Anthony and *Coffea salvatrix* Swynn. & Philipson presented higher concentration of mangiferin than *C. arabica*, *Coffea eugenoides*, *Coffea heterocalyx* Stoff., *Coffea pseudozanguebariae*, or *C. sessiliflora.* Campa et al. studied the mangiferin and hydroxycinnamic acid ester content in 23 *Coffea* leaves. Their histochemical observations revealed that mangiferin was present in low concentration in *C. arabica* mainly in the exocarp, mesocarp, and fruits [30, 33]. Alves et al. determined that, even though 90% of tocopherols content remains unchanged after the roasting process of Arabica and Robusta coffee seeds, the concentration of β-tocopherol is reduced by 25% in Robusta coffee, aspect which can be used as a discrimination tool between the two *Coffee* species [34].

4. Metabolism of caffeine in *Coffea arabica* L.

4.1. Biosynthesis of caffeine

Caffeine biosynthesis takes place in the upper leaves and in the pericarp, and it is absent in the second and third leaves, cotyledons, lower stem, and root. After the biosynthesis, caffeine is accumulated in the mature leaves of coffee, but when the seed starts growing inside the fruit, it is translocated through the membranes being accumulated in the endosperm. The final quantity of caffeine is reached in 8 months after flowering [10, 35]. *C. arabica* leaves have the highest caffeine content, meanwhile *C. salvatrix*, *C. eugenioides*, and *C. bengalensis* leaves contain three to seven times lower concentrations [36].

The caffeine biosynthesis in leaves is age-dependent which means that it occurs usually at the very early stages of the leaf development reaching a maximum when the leaves are fully opened. The same age-dependent biosynthesis was observed in the fruits of coffee and in the leaves, flowers, and fruits of tea plants. Naoko and Hiroshi studied the levels of purine alkaloids and the metabolism of adenine in the first and in the second leaves from the shoot apices of coffee plants. Even though theobromine and caffeine were found in these leaves, theophylline was not detected. Studies showed that adenine was converted to theobromine and caffeine in the first leaves and the degradation of adenine nucleotides was low in both types of leaves [18].

Caffeine is 1, 3, 7-trimethylxanthine having a xanthine skeleton derived from purine nucleotides. The purine compound initially involved in the biosynthesis pathway of caffeine is xanthosine, which is a substrate for the methyl group donated by *S*-adenosyl methionine (SAM). The most important pathway for caffeine biosynthesis, which was proposed by Hiroshi and Thomas, is as follows: xanthosine → 7-methylxanthosine →7-methylxanthine → theobromine → caffeine. The first methylation step is the conversion of xanthosine to 7-methylxanthosine being catalyzed by 7-methylxanthosine synthase (an N-methyltransferase). The next step of 7-methylxanthosine hydrolysis to 7-methylxanthine is catalyzed by methylxanthine nucleosidase. The conversions of 7-methylxanthine to theobromine and then theobromine to caffeine are catalyzed by N-methyltransferases (firstly identified 26 years ago by Takeo Suzuki and Ei-ichi Takahashi). The caffeine synthase is a monomeric enzyme with an optimum pH of 8.5. This enzyme is not inhibited by caffeine, but instead, there is a complete inhibition by low concentrations of S-adenosyl-L-homocysteine (SAH), therefore its activity is regulated by SAM:SAH ratio. Caffeine synthase is found in chloroplast, but it is not affected by light, thus the caffeine synthesis takes place also in darkness [35, 37]. By summing up, the N-methyltransferases, which are involved in caffeine biosynthesis pathway, are: 7-methylxanthine 3-N-methyltransferase, caffeine xanthine methyltransferase 1, caffeine methylxanthine methyltransferase 2, and caffeine dimethylxanthine methyltransferase [38].

The caffeine synthesis is affected mostly by the enzymes activity, which can appear as limiting factors in the biosynthesis: xanthosine N-methyltransferase, 7-methylxanthosine nucleosidase, 7-methylxanthine N-methyltransferase, and theobromine N-methyltransferase. The observation from the tea leaves indicated that their activity is also affected by seasons [18]. Caffeine accumulation seems to be regulated by the genes, which encode N-methyltransferase and caffeine (7-N) demethylase [36].

4.2. Catabolism of caffeine

The catabolism of caffeine is a slow process, which begins with the removal of the three methyl groups from the skeleton resulting in xanthine which is further decomposed to CO_2 and NH_3. These degradation reactions are catalyzed by various demethylases that have different levels of activity in the *Coffee* species (higher in *C. eugenioides* than in *C. arabica*). The major rate-limiting step in caffeine catabolism is the caffeine conversion into theophylline [35]. Catabolism pathways involve the conversion of caffeine into theophylline, 3-methylxanthine, xanthine, uric acid, allantoin, allantoic acid, urea and finally results in $CO_2 + NH_3$ [36].

Even though the degradation of caffeine is negligible in the leaves of *C. arabica, C. salvatrix,* and *C. bengalensis,* the leaves of *C. eugenioides* with low caffeine content metabolize caffeine rapidly by degradation to CO_2 within 24 hours. The explication could be that this taxon contains higher levels of N-7-demethylase than *C. arabica*; therefore, caffeine is efficiently converted to theophylline being quickly metabolized afterwards [35, 36].

The catabolism of caffeine can also be achieved by various bacteria like *Pseudomonas cepacia, Pseudomonas putida,* and *Serratia marcescens.* Bacterial degradation is different from other pathways since caffeine is transformed to theobromine, after that to 7-methylxanthine, xanthine, and finally to NH_3. The conversion of caffeine to theobromine is catalyzed by N-1-demethylase, which was successfully isolated from *Pseudomonas putida* [35].

Some studies elaborated by Suzuki also showed that theophylline and xanthine were more quickly metabolized in the immature fruit; meanwhile, the caffeine degradation was performed in both mature and immature coffee fruits. The results showed that adenine was a more effective precursor than guanine and L-methyl-methionine for the biosynthesis of N7-methylxanthine, theobromine, and caffeine. These results underlined that the caffeine biosynthesis occurred mostly in immature fruits through methylation of N-methylxanthine and theobromine; meanwhile, its biodegradation occurred through theophylline, which is accumulated after that the seed is full size and the fruit is mature [39].

5. Analytical methods used for caffeine determination

Caffeine is a deeply studied alkaloid and, as it presents a significant importance for science and human body, many analytical methods have been developed over the years for its determination. The aim of these analytical methods was to identify and quantify this compound with different origin (from plants, beverages, and medicines). The most important and the most used techniques are the chromatographic methods coupled with spectrometry which allow a qualitative and quantitative determination of caffeine. Since these methods are quite expensive, there are a lot of scientific reports about different new low cost techniques. Electroanalytical methods recently became more popular since they are faster, more convenient, present lower costs, and are environmentally friendly in comparison with the other conventional analytical methods.

5.1. Chromatographic methods

The biochemical diversity of wild accessions of *C. arabica* (38 genotypes) and *C. canephora* (38 genotypes) was analyzed by high performance liquid chromatography (HPLC) method by Ky et al. The analyzed compounds responsible for coffee aroma were caffeine, trigonelline, chlorogenic acids, and sucrose. Sucrose was analyzed using anion-exchange chromatography coupled to pulsed amperometric detection. An aqueous solution (containing triethylamine and acetic acid) and methanol were used as mobile phases for caffeine and trigonelline, meanwhile their ultraviolet (UV) detection was carried out at 272.8 nm (maximum absorption of caffeine) and 263.3 nm (trigonelline maximum absorption) wavelength. *C. arabica* contained more trigonelline and sucrose, meanwhile *C. canephora* presented higher concentration of chlorogenic acids and caffeine. The results underlined that *C. canephora* has a higher compounds diversity than *C. arabica* excepting trigonelline and sucrose. In addition, there were not identified differences for alkaloids and sucrose between *C. canephora* accessions [40].

Mazwfera et al. used HPLC method for qualitative and quantitative determinations of caffeine, theobromine, and theophylline in aqueous extracts of endosperm from immature and mature fruits of *C. arabica, C. canephora, C. benghalensis, C. dewevrei, C. eugenioides, C. stenophylla,* and *C. salvatrix*. The highest concentration of caffeine was found in *C. canephora* in both immature and mature endosperms; meanwhile, caffeine was not detected in extracts of *C. bengalensis* mature fruit. Moreover, caffeine was more slowly metabolized by *C. arabica* and *C. canephora* immature endosperm than the other five species [41].

Nowadays, science also pays attention for low-pressure chromatography, which presents advantages such as easier assembly and handling, lower implementation cost, and more widespread application regarding separation of neutral or ionic compounds than other separation techniques. By using this method, caffeine was determined from six different coffee beverages: three regular, one decaffeinated, one soluble, and one chicory blended coffees. The results showed similar values for caffeine compared with the reference method (HPLC-UV) with 5% relative deviations, with no signal for theobromine or theophylline. Even though low-pressure chromatography uses a short column which implies reduced resolution, this methodology can be a competitive alternative to usual HPLC for caffeine determination in different materials due to its advantages such as higher determination rate, lower consumption of reagents, and no need to degas the mobile phase [42].

5.2. Spectroscopic methods

Frizzarin used the dispersive liquid-liquid microextraction and spectrophotometric determination for caffeine in different coffee beverages. This lab-in-syringe sequential injection analysis system is a fast and simple procedure, which uses dichloromethane with high extraction capacity and good selectivity and methanol as dispersing agent. The developed technique presented linear response range from 2 to 75 mgL^{-1}, limit of detection of 0.46 mgL^{-1} and limit of quantification of 1.54 mgL^{-1}, being successfully applied for caffeine determination from brewed, instant, and decaffeinated coffee samples [43].

For the determination of caffeine, formic acid, trigonelline, and 5-(hydroxyl-methyl) furfural, there was applied the proton nuclear magnetic resonance technique (1H NMR) without any derivatization. The limits of detection were 1.32 mg/g for caffeine, 0.58 mg/g for trigonelline, and 0.30 mg/g for 5-(hydroxyl-methyl) furfural. In addition, HPLC was also used for the determination of these compounds employing a diode-array detector at 273 nm for caffeine, 265 nm for trigonelline, and 284 nm for 5-(hydroxyl-methyl) furfural [44].

Alesso used spectrofluorimetric method for caffeine determination, which was developed by using the quenching effect on fluorescent emission of bovine serum albumin molecule (λ_{em} = 338 nm, λ_{ex} = 280 nm). During the investigation, the sampling rate was increased to 60 samples/hour using potassium dihydrophosphate buffer (pH 6.8) as carrier with a flow rate of 1.5 $mLmin^{-1}$. Several parameters were optimized such as the nature and the concentration of both the buffer and the fluorophore, and the carrier flow rate, leading to a linear range from 6.68×10^{-6} to 4.0×10^{-3} $molL^{-1}$. This sensitive and selective method was successfully employed for caffeine determination from various matrices such as energy drinks, dietary supplements, and slimming infusion samples without any pretreatment [45].

A comparative study about caffeine content in roasted ground coffee and in China black tea was performed using a liquid-liquid extraction. A simple and rapid spectrophotometric method was described for the determination of caffeine indicating that caffeine concentration in roasted coffee is lower than its concentration in black tea [46].

5.3. Electrochemical methods

Rotko and Beczkowska elaborated a Nafion covered lead film sensor leading to a sensitive, selective, and low cost method applied for caffeine determination in tea, coffee, soft and energy drink samples, and pharmaceutical products. Two anodic peaks were detected at 0.86 and 1.40 V (versus Ag/AgCl) in acidic medium, and the corresponding detection limits were equal to 1.7×10^{-8} and 2.2×10^{-7} $molL^{-1}$, respectively, at 120 seconds of accumulation time. The results were in good agreement with the concentrations mentioned by the manufacturer and with those reported by using the spectrophotometric method [47].

Similar studies were reported by Gao et al. who modified a glassy carbon electrode (GCE) with large mesoporous carbon and Nafion composites (LMC/Nafion) achieving a simultaneous determination of theophylline and caffeine in serum and beverages. Cyclic voltammetry (CV) and differential pulse voltammetry (DPV) were employed for the investigation of the electrochemical behaviors of theophylline and caffeine on the LMC/Nafion/GCE. Simultaneous determination of these two alkaloids was successfully obtained at the modified electrode since peak-to-peak separation of the two DPV peaks was about 150 mV. The sensor performances included detection limits of theophylline and caffeine of 0.37 and 0.47 μM, and a linear range between 0.8–180.0 and 1.3–230.0 μM, respectively [48].

The boron doped diamond electrode was also successfully used for simultaneous determination of caffeine and chlorogenic acid by CV and adsorptive stripping voltammetry, applied on commercial beverages. Various experimental parameters such as dependence of peak current and potential on pH, scan rate, and accumulation time were optimized. The oxidation peak potentials of caffeine and chlorogenic acid from binary mixtures were separated with 0.4 V in acidic medium by employing square-wave stripping voltammetry. The analytical performances of this sensor indicated detection limits of 0.107 $mgmL^{-1}$ ($5.51 \cdot 10^{-7}$ M) for caffeine and of 0.448 $mgmL^{-1}$ (1.26×10^{-6} M) for chlorogenic acid [49].

Samanidou reviewed in her work several electroanalytical methods for caffeine determination. One example was constituted by polymer-modified glassy carbon electrode, which was electropolymerized with 4-amino-3-hydroxynaphthalene sulfonic acid. This sensor was suitable for a high sensitive, selective, and stable caffeine determination in coffee without any interference. The polymer-modified electrode presented a linear range of 6×10^{-8}–4×10^{-5} $molL^{-1}$ and a detection limit of 1.37×10^{-7} $molL^{-1}$ [44]. Other authors employed 1,4-benzoquinone modified carbon paste electrode for indirect voltammetric determination of caffeine by using square wave voltammetry (SWV) and CV with detections limits of 0.3 and 5.1 $\mu molL^{-1}$, respectively [44].

Fritea et al. studied the formation of inclusion complexes between β-cyclodextrin (β-CD) and caffeine. The relationship between the oxidation peak currents and the concentration of caffeine in the presence of β-CD was examined by SWV indicating that the molar ratio of 1:1 is convenient for complexation [50]. Voltammetric methods were successfully applied for the simultaneous determination of some alkaloids (caffeine, aminophylline, theophylline, codeine phosphate, and papaverine hydrochloride) in different pharmaceutical combinations and in urine samples by using an electrochemically activated GCE [51, 52].

6. Medicinal importance of coffee and caffeine

6.1. Ethnomedicinal knowledge of coffee

There are several references related to the application of coffee plants in traditional medicine underlying multiple healing potentials. In addition to the official drug (*Coffeae semen*) utilization, there are data about the use of other plant parts in the Equator region. These coffee plant parts were used to treat various diseases in both human and veterinary medicines.

Ross presented an ethnomedicinal map about the use of *Coffee* species (different plant parts and different routes of administration) as a medicine throughout the world describing a wide range of diseases or symptoms, such as diarrhea, intestinal pain, HIV/AIDS, flu, anemia, edema, asthenia, liver diseases, migraine, stomach pain, fever; against bleeding that accompanied abortion; as astringent, aphrodisiac, cough suppressant; for cardiotonic and neurotonic effects, for tiredness, asthma, scorpion bites, and for the production of prolactin [2, 53–55].

In traditional medicine, the coffee coal was used for the inflammatory diseases of mouth and pharynx, but also it was used as a treatment of festering wounds [56]. In Nepal, *C. benghalensis* flowers were used as a treatment for excessive bleeding during menstruation [2, 57]. Different plant parts of *C. canephora* were used for backache, measles, coughing, and jaundice [2, 58].

There are information about utilization of coffee leaves and seeds as infusion or decoction having different regional names such as "giser" in Yemen [2, 12]. Native people from Ethiopia drank a beverage named "hoja" for diarrhea and nausea, which were caused by poisoning [2, 59]. In some regions of Indonesia and Ethiopia, people used to prepare a tea from *C. arabica* or *C. robusta* leaves, named "copi daon" or "leaf coffee" [2, 60]. In Liberia, women used to prepare a coffee leaf infusion for their children. The infusion from *C. arabica* leaves was tested on the London markets, but people said that it was undrinkable [2, 60].

Schmid et al. presented a study about the veterinary use of coffee in Swiss provinces indicating that farmers used coffee as a beverage to treat the reproductive, gastrointestinal, and metabolic disorders and infertility in animals. The authors showed that a subcutaneous injection of 10 mL coffee seed extract increased the healing rate of the newborn calves from diarrhea in 30% of the cases compared with the control subjects [2, 61].

6.2. Medicinal importance of coffee

The seed extracts of *Coffea* species have presented several health benefits due to the polyphenols content being used in cosmetics and pharmaceutical industry. The pharmacological benefits included a wide range of effects such as antioxidant, detoxifying, lipid reducing, cardioprotective, anti-inflammatory, analgesic, antineoplastic, diuretic, antibacterial, antiviral, antifungal, antiosteoporotic, anticelulitic, and anti-age activity, the effect on central nervous and gastrointestinal systems, and on blood vessels [62]. According to Rodriguez et al., the hydroalcoholic extract of coffee silverskin can be used for topical application because it has no irritant effects. Three different extracts were studied in this case performing *in vitro* and *in vivo* skin and ocular irritation assays [2, 63].

Nowadays, many decaffeinated beverages are produced in order to overcome the negative withdrawal effects of caffeine by using different extraction methods, which involve some toxic solvents. Due to these methods, they may be harmful for the human body; therefore, many studies were performed in order to obtain new and less toxic extraction methods. Among these new techniques, it can be mentioned: the microbiological caffeine degradation by using *Pseudomonas* and *Aspergillus* strains, enzymatic caffeine removal, and the genetically reduction of caffeine in plant [2, 64]. Even though many new decaffeination methods were successfully tested, the decaffeinated coffee still contains a minimum quantity of caffeine, aspect which has to be taken into consideration by patients vulnerable to caffeine effects. McCusker et al. have evaluated the caffeine concentration in various decaffeinated coffee drinks collected from different sources obtaining a caffeine content of 0–13.9 and 12.0–13.4 mg/16-oz serving [65].

Ross mentioned in his work many scientific tests about the topical utilization of green seed extracts. The results showed that these extracts have a significant anti-inflammatory activity. Other tests made on mice showed that the extracts of dried seeds have anticancer effect and they can also decrease blood sugar levels. However, seven cups of coffee per day with alcohol and cigarette can increase the suicidal tendency and the gastric acid level [2, 53]. The regular consumption of coffee mainly reduces the occurrence of kidney and liver cancers; meanwhile, premenopausal, breast, and colon cancers are less influenced. These effects are attributed to the content in caffeine, diterpenoids, caffeic acid, polyphenols, essential oils, and heterocyclic molecules [2, 66].

Cooper and Kronenberg tested the topical effect of coffee seed extract on 30 patients with dermatological problems. During this study, the product was applied on the whole facial area of 20 patients. In the case of 10 patients, it was applied only on the half of their face and the remaining area was treated with a placebo cream. The investigated cream presented noticeable effects such as appearance of fine lines and reduction of wrinkles and pigmentation [2, 67].

Jessen et al. have proved that thermogenic effect of nicotine can be improved by caffeine. They have studied this effect by using chewing gums with different concentration of nicotine and caffeine. They have found that the thermogenic effect of 1 mg of nicotine can be doubled by 100 mg of caffeine; therefore, caffeine could be used for weight gain prevention after smoking cessation [68]. It was also demonstrated that regular consumption of coffee presented benefits for the respiratory system in patients with asthma and smokers. In addition, the regular coffee consumption can reduce the occurrence probability of type II diabetes by 60%. The effect was not generated by caffeine since both caffeinated and decaffeinated beverages were studied obtaining the same result. Therefore, the compounds which are responsible for this effect are still unknown. Due to this reduced risk for diabetes development, coffee may be indicated as "functional food" efficient for the metabolic diseases prevention. Coffee presents high antioxidant properties due to the presence of flavonoids and polyphenols in its composition. The antioxidant activity was significantly increased after consumption of unfiltered coffee because of the glutathione increase [2, 69, 70].

Wagemaker et al. characterized the lipid fraction of the coffee seeds determining the variable concentration of wax, oil, and unsaponifiable material in the case of 10 *Coffea* species. The sunscreen effect was calculated between 0.0 and 4.1 SPF depending on the species indicating that the presence of linoleic and oleic acids is very useful for cosmetic products [2, 71]. Phenolic acids presented inhibitory effect on skin tumor in mice [2]. The influence of 13 phenolic acids on the phenol sulfotransferase enzyme activity was investigated since this enzyme is involved in the detoxification process. Some acids have inhibited the enzyme by 21–30% (chlorogenic, syringic, protocatechuic, vanillic, sinapic, and caffeic acids), while other acids have enhanced its activity (p-hydroxybenzoic, gallic, gentisic, o-coumaric, p-coumaric, m-coumaric, and ferulic acids) [23].

Other studies reported that isoquercitrin and rutoside had anti-atherosclerotic effect, quercitrin had positive chronotropic, positive inotropic, and anti-arrhythmic effects, which were tested on guinea pigs. Both quercetin and rutoside are used as therapeutic agents for capillary fragility and phlebosclerosis [2, 22]. Some flavonoids such as rutoside, gossypin, naringenin, (+)-Cyanidanol-3, quercetin, kaempferol, and rutin exhibit antiulcer activity by enhancing the gastric protection conferred by mucous content and inhibiting the platelet activating factor. Quercetin, kaempferol, and rutin showed antidiarrhetic effect and also benefits in other intestinal diseases since their actions are mediated through α2-adrenergic and calcium systems. Some flavonoids such as rutin and venorutin were also proved to exhibit hepatoprotective effects. In addition, kaempferol, rutoside, and quercetin showed antioxidant, antiviral, antifungal, antibacterial, anti-inflammatory, and antiallergic activity [22].

In one of our previous studies, the antioxidant activity of three *Coffea* species was tested using enhanced chemiluminescence (ECL), 2,2-diphenyl-1-picrylhydrazyl (DPPH), and oxygen radical absorbance capacity (ORAC). Even though trolox equivalent values obtained by ECL and DPPH methods showed loose correlation, values obtained by ORAC assay were higher without correlation in each plant. The differences in the reactive antioxidant compounds among the assays and the altered reactivity with the reporter molecules might be responsible for this much higher antioxidant activities measured by the ORAC assay. However, a closer correlation was observed between the ECL method and the scavenging potential of the DPPH technique in each species, much higher DPPH antioxidant activity was observed at the immature pericarp of *C. benghalensis* than in case of the other species [29].

The hepatoprotective effect of coffee brews was proven by Lima et al. The two enzymes (aspartate aminotransferase and alanine aminotransferase)that are relevant for liver damage, the thiobarbituric acid reactive species, and total lipids, all were decreased. This effect was not negatively influenced by decaffeination process; meanwhile, it was indicated that the roasted coffee brews had a better protection against liver disease compared with green coffee brews [2, 72]. The vascular effects of coffee polyphenols were investigated by Ochiai et al., and the results showed that the peripheral endothelial function was improved after glucose loading in healthy person due to the ingestion of coffee polyphenols; meanwhile, no radical changes of the antioxidant activity were noticed [2, 73]. Chandra successfully assessed the *in-vitro* anti-inflammatory effect of the *C. arabica* aqueous extract with different concentrations against the denaturation of protein by incubation with egg albumin. The anti-inflammatory activity of the coffee extract was related to the polyphenols content and it was higher in comparison with the effect of diclofenac sodium [2, 74].

6.3. Medicinal importance of caffeine

Caffeine is absorbed 99% from stomach (20%) and small intestine (80%) [75]. The most popular effect of caffeine is its stimulant effect on the central nervous system. Moreover, there is a wide range of various effects that are also attributed to caffeine such as increase of the painkillers effect, reduction of tiredness and of migraine (as it has a vasoconstrictor activity in the brain), increase of the stomach acid secretion, stimulation of the heart function (with hypertensive effects), stimulation of kidney function (with diuretic effect), stimulation of respiration, and decrease of the vitamin B concentration [2]. In addition, it is supposed that the occurrence of Parkinson's disease may be reduced by a regular consumption of coffee and cola [2].

Armanian showed that this alkaloid presents significant preventive effect of apnea in premature infants. Only 15.4% of the infants treated with caffeine have developed apnea in comparison with 61.5% of the control group [76].

As it was mentioned before, caffeine has a lot of beneficial effects, but in excess, it can generate plenty of secondary effects on the gastrointestinal, central nervous, circulating, and respiratory systems such as intestinal irritability, vomiting, diarrhea, stomach ulcers, tremors, sleep disturbances, headache, hallucinations, epileptic convulsions, high blood pressure, arrhythmic tachycardia, numbness, muscle spasms, and respiratory paralysis [2]. The excessive coffee consumption can lead to caffeinism which is an addiction because caffeine stimulates the central nervous system and also possesses negative withdrawal effects. Due to those secondary effects, coffee consumption should be avoided by people having cardio-vascular, kidney, neurological and gastric diseases, hyperthyroidism, and caffeine sensitivity [2]. In addition, caffeine is not recommended during pregnancy and lactation as it can cause spontaneous abortion and it can be secreted in the breast milk [2].

7. Conclusions

Based on this comprehensive work, we can conclude that *C. arabica* is the most studied and used plant both in scientific and social fields from the past up to the present. This plant presents a rich content of caffeine, a purine alkaloid, which is the most popular and well-known compound from the coffee plant. This substance showed important physiological effects on human body, but also including some serious secondary effects. Moreover, the scientific research is focused mostly on the physiological effects of coffees, but both the metabolism and extraction methods of caffeine have to be well studied in order to achieve a safer use of caffeine.

Author details

Eva Brigitta Patay[1], Luminiţa Fritea[2], Andreea Antonescu[2], Angela Antonescu[2] and Luciana Dobjanschi[2]*

*Address all correspondence to: dobjanschil@yahoo.com

1 Department of Pharmacognosy, Faculty of Pharmacy, University of Pécs, Pécs, Hungary

2 Department of Preclinical Disciplines, Faculty of Medicine and Pharmacy, University of Oradea, Oradea, Romania

References

[1] Davis AP, Chester M, Maurin O, Fay MF. Searching for the relatives of Coffea (Rubiaceae, Ixoroidae): The circumscription and phylogeny of Coffeae based on plastid sequence data and morphology. American Journal of Botany. 2007;**94**:313–329

[2] Patay ÉB, Bencsik T, Papp N. Phytochemical overview and medicinal importance of Coffea species from the past until now. Asian Pacific Journal of Tropical Medicine. 2016;**9**(12):1127–1135

[3] Tesfaye K, Govers K, Bekele E, Borsch T. ISSR fingerprinting of *Coffea arabica* throughout Ethiopia reveals high variability in wild populations and distinguishes them from landraces. Plant Systematics and Evolution. 2013;**300**(5):881–897

[4] Rohwer JG. A Trópusok Növényei. Budapest: Magyar Könyvklub; 2002. p. 148

[5] Crozier A, Ashihara H, Tomás BF. Teas, cocoa and coffee. In: Plant Secondary Metabolites and Health. Chichester: Blackwell Publishing; 2012. pp. 4–5

[6] Barone JJ, Roberts HC. IV Human Consumption of Caffeine. Berlin: Springer; 1984. pp. 59–73

[7] Rácz J. Növénynevek Enciklopédiája. Az Elnevezések Eredete, A Növények Kultúrtörténete És Élettani Hatása. Budapest: Tinta Könyvkiadó; 2010. pp. 393–395

[8] Baeza G, Benavent MA, Sarriá B, Goya L, Mateos R, Bravo L. Green coffee hydroxycinnamic acids but not caffeine protect human HepG2 cells against oxidative stress. Food Research International. 2014;**62**:1038–1046

[9] Brücher H. Tropische Nutzpflanzen. Berlin: Springer; 1977. pp. 459–468

[10] Janzen SO. Chemistry of Coffee. Hamburg: Elsevier; 2013. pp. 1085–1113

[11] Măndiţă D. Ce ştim şi ce nu ştim despre cafea. Bucureşti: Editura Tehnică. 2008;**37**:9–14

[12] Nowak B, Schulz B. A Trópusok Gyümölcsei. Budapest: Magyar Könyvklub; 2002. pp. 165–166

[13] Patay ÉB, Nemeth T, Nemeth TS, Papp N. Coffea taxonok biológiai, fitokémiai és gyógyászati értékelése. Botanikai Közlemények. 2014;**101**(1–2):263–280

[14] Buffo RA, Freire CC. Coffee flavour: An overview. Flavour and Fragrance Journal. 2004;**19**:99–104

[15] Farah A. Coffee constituents. In: Coffee: Emerging Health Effects and Disease Prevention. 1st ed. John Wiley & Sons; New Jersey 2012. pp. 21–58

[16] Fattorusso E, Scafati OT. Modern Alkaloids. Structure, Isolation, Synthesis and Biology. Weinheim: Wiley; 2008. p. 58

[17] Friedman J, Waller GR. Caffeine hazards and their prevention in germinating seeds of coffee (*Coffea arabica* L). Journal of Chemical Ecology. 1983;**9**(8):1099–1106

[18] Fujimon N, Ashira H. Biosinthesis of theobromine and caffeine in developing leaves of *Coffea arabica*. Pkytochnnrury. 1994;**36(**6):1359–1361

[19] Kretschmar JA, Baumann TW. Caffeine in citrus flowers. Phytochemistry. 1999;**52**:19–23

[20] Huang MT, Smart RC, Wong CQ, Conney AH. Inhibitory effect of curcumin, chlorogenic acid, caffeic acid, and ferulic acid on tumor promotion in mouse skin by 12-*O*-Tetradecanoylphorbol-13-acetate. Cancer Research. **1988;48**:5941–5946

[21] Mussatto SI, Ballesteros LF, Martins S, Teixeira JA. Extraction of antioxidant phenolic compounds from spent coffee grounds. Separation and Purification Technology. 2011;**83**:173–179

[22] Narayana KR, Reddy MS, Chaluvadi MR, Krishna DR. Bioflavonoids classification, pharmacological, biochemical effects and therapeutic potential. The Indian Journal of Pharmacology. 2001;**33**:2–16

[23] Yeh **CT,** Yen GC. Effects of phenolic acids on human phenolsulfotransferases in relation to their antioxidant activity. The Journal of Agricultural and Food Chemistry. **2003;51:**1474–1479

[24] Naidu MM, Sulochanamma G, Sampathu SR, Srinivas P. Studies on extraction and antioxidant potential of green coffee. Food Chemistry. 2008;**107**:377–384

[25] Dziki UG, Świeca M, Dziki D, Kowalska I, Pecio L, Durak A, Seczyk L. Lipoxygenase inhibitors and antioxidants fromgreen coffee – mechanism of action in the light of potential bioaccessibility. Food Research International. 2014;**61**:48–55

[26] Gîrd CE, Nencu I, Costea T, Duțu LE, Popescu ML, Ciupitu N. Quantitative analysis of phenolic compounds from *Salvia officinalis* L. Leaves. Farmacia. 2014;**62**:649–657

[27] Quiroz MLS, Campos AA, Alfaro GV, Rios OG, Villeneuve P, Espinoza MCF. Isolation of green coffee chlorogenic acids using activated carbon. The Journal of Food Composition and Analysis. 2014;**33**:55–58

[28] Patay ÉB, Németh T, Németh TS, Filep R, Vlase L, Papp N. Histological and phytochemical studies of *Coffea benghalensis* B. Heyne Ex Schult, compared with *Coffea arabica* L. Farmacia. 2016;**64**(1):125–130

[29] Patay ÉB, Sali N, Koszegi T, Csepregi R, Balázs VL, Németh TS, Németh T, Papp N. Antioxidant potential, tannin and polyphenol contents of seed and pericarp of three Coffea species. Asian Pacific Journal of Tropical Medicine. 2016;**9**(4):366–371

[30] Campa C, Mondolot L, Rakotondravao A, Bidel LPR, Gargadennec A, Couturon E, Fisca PL, Rakotomalala JJ, Allemand CJ, Davis AP. A survey of mangiferin and hydroxycinnamic acid ester accumulation in coffee (Coffea) leaves: Biological implications and uses. Annals of Botany – London. 2012;**110**:595–613

[31] Talamond P, Mondolot L, Gargadennec A, Kochko A, Hamon S, Fruchier A, Campa C. First report on mangiferin (C-glucosyl-xanthone) isolated from leaves of a wild coffee plant, *Coffea pseudozanguebariae* (Rubiaceae). Acta Botanica Gallica. 2008;**155**:513–519

[32] Mondolot L, Fisca PL, Buatois B, Talansier E, Kocho AD, Campa C. Evolution in caffeoylquinic acid content and histolocalization during *Coffea canephora* leaf development. Annals of Botany. 2006;**98**(1):33–40

[33] Farah A, Donangelo CM. Phenolic compounds in coffee. The Brazilian Journal of Plant Physiology. 2006;**18**(1):23–36

[34] Alves RC, Casal S, Alves MR, Oliveira MB. Discrimination between arabica and robusta coffee species on the basis of their tocopherol profiles. Food Chemistry. 2009;**114**:295–299

[35] Ashihara H, Crozier A. Caffeine: A well known but little mentioned compound in plant science. Trends in Plant Science. 2001;**6**(9):407–413.

[36] Ashihara H, Crozier A. Biosynthesis and catabolism of caffeine in low-caffeine-containing species of coffea. Journal of Agriculturaland Food Chemistry. 1999;**47**:3425–3431

[37] Baumann TW, Laser ED, Wanner H. 7-methylxanthosine – an intermediate in caffeine biosynthesis. Phytochemistry. 1978;**17**(12):2075–2076

[38] Aniszewski T. Enzymes specifically involved in alkaloid biosynthesis. In: Alkaloid Chemistry, Biological Significance, Applications, and Ecological Role. Amsterdam: Elsevier; 2007. p. 176

[39] Suzuki T, Waller GR. Biosynthesis and biodegradation of caffeine, theobromine, and theophylline in *Coffea arabica* L. Fruits. Journal of Agrlcultural and Food Chemistry. 1984;**32**:845–848

[40] Ky CL, Louarn J, Dussert S, Guyot B, Hamon S, Noiro M. Caffeine, trigonelline, chlorogenic acids and sucrose diversity in wild *Coffea arabica* L. and *C. canephora* P. accessions. Food Chemistry. 2001;**75**:223–230

[41] Mazwfera P, Crozier A, Magalhae AC. Caffeine metabolism in *Coffea arabica* and other species of coffee. Phytochemistry. 1991;**30**(12):3913–3916

[42] Santos JR, Rangel AOSS. Determination of caffeine in coffee using low-pressure chromatography. In: Coffee in Health and Disease Prevention. London: Elsevier; 2015. pp. 983–991

[43] Frizzarin RM, Maya F, Estela JM, Cerdà V. Fully-automated in-syringe dispersive liquid-liquid microextraction for the determination of caffeine in coffee beverages. Food Chemistry. 2016;**212**:759–767

[44] Samanidou VF. Determination of polyphenols and major purine alkaloids in coffee: An overview. In: Coffee in Health and Disease Prevention. London: Elsevier; 2015. pp. 971–981

[45] Alesso M, Fernandez LP. Caffeine determination by flow injection analysis employing bovine serum albumin as a fluorophore. Microchem Journal. 2016;**127**:165–169

[46] Purcărea C, Chis A, Vicas S, Fodor A. Comparative studies about caffeine content in roasted ground coffee and in China black tea. Analele Universitatii din Oradea. 2008;**7**:966–971

[47] Rotko KT, Beczkowska I. Nafion covered lead film electrode for the voltammetric determination of caffeine in beverage samples and pharmaceutical formulations. Food Chemistry. 2015;**172**:24–29

[48] Gao Y, Wang H, Guo L. Simultaneous determination of theophylline and caffeine by large mesoporous carbon/nafion modified electrode. Journal of Electroanalytical Chemistry. 2013;**706**:7–12

[49] Yardım Y, Keskin E, Sentürk Z. Voltammetric determination of mixtures of caffeine and chlorogenic acid in beverage samples using a boron-doped diamond electrode. Talanta. 2013;**116**:1010–1017

[50] Fritea L, Tertis M, Topală T, Săndulescu R. Electrochemical and spectral study of cyclodextrins interactions with some pharmaceutical substances. Farmacia. 2013;**61**(6):1054–1068

[51] Câmpean A, Tertis M, Săndulescu R. Electrochemical behavior of some purine derivatives on carbon based electrodes. Central European Journal of Chemistry. 2011;**9**(3):466–473

[52] Câmpean A, Tertis M, Săndulescu R. Voltammetric determination of some alkaloids and other compounds in pharmaceuticals and urine using an electrochemically activated glassy carbon electrode. Central European Journal of Chemistry. 2011;**9**(4):688–700

[53] Ross IA. Medicinal Plants of the World. 3rd ed. New Jersey: Humana Press Inc; 2005. pp. 155–184

[54] Lamorde M, Tabuti JRS, Obua C, Collins KB, Lanyero H, Pauline BK, Bbosa GS, Lubega A, Okeng JO, Ryan M, Waako PJ, Concepta M. Medicinal plants used by traditional medicine practitioners for the treatment of HIV/AIDS and related conditions in Uganda. Journal of Ethnopharmacology. 2010;**130**:43–53

[55] Neuwinger HD. African Traditional Medicine. Stuttgart: Medpharm Scientific Publishers; 2000. p. 130

[56] Gruenwald J, Brendler T, Jaenicke C. PDR for Herbal Medicines. Montvale: Medical Economics Company; 2000. pp. 202–204

[57] Ghimire K, Bastakoti RR. Ethnomedicinal knowledge and healthcare practices among the Tharus of Nawalparasi district in central Nepal. Forest Ecology Management. 2009;**257**:2066–2072

[58] Tabuti JRS, Lye KA, Dhillion SS. Traditional herbal drugs of Bulamogi, Uganda: Plants, use and administration. Journal of Ethnopharmacology. 2003;**88**:19–44

[59] Belayneh A, Bussa NF. Ethnomedicinal plants used to treat human aliments in the prehistoric place of Harla and Dengego valleys, eastern Ethiopia. Journal of Ethnobiology and Ethnomedicine. 2014;**10**:18

[60] Cramer PJS. A Review of Literature of Coffee Research in Indonesia. Costa Rica: Miscellaneous publication; 1957. p. 175

[61] Schmid K, Ivemeyer S, Vogl C, Klarer F, Meier B, Hamburger M, Walkenhorst M. Traditional use of herbal remedies in livestock by farmers in 3 swiss cantons (Aargau, Zurich, Schaffhausen). Forsch Komplement Medicine. 2012;**19**:125–136

[62] Boros B, Jakabová S, Dörnyei A, Horváth GY, Pluhár ZS, Killár F, Felinger A. Determination of polyphenolic compounds by liquid chromatography-mass spectrometry in thymus species. Journal of Chromatography A. 2010;**1217**:7972–7980

[63] Rodrigues F, Pereira RC, Pimentel FB, Alves RC, Ferreira M, Sarmento B, Amaral MH, Oliveira MBPP. Are coffee silverskin extracts safe for topical use? An in vitro and in vivo approach. Industrail Crops and Products. 2015;**63**:167–174

[64] Gokulakrishnan S, Chandraraj K, Gummadi SN. Microbial and enzymatic methods for the removal of caffeine. Enzyme and Microbiology Technology. 2005;**37**:225–232

[65] McCusker RR, Fuehrlein B, Goldberger BA, Gold MS, Cone EI. Caffeine content of decaffeinated coffee. Journal of Analytical Toxicology. 2006;**30**:611–613

[66] Nkondjock A. Coffee consumption and the risk of cancer. An overview. Cancer Letters. 2009;**277**:121–125

[67] Cooper R, Kronenberg F. Botanical Medicine. New Rochelle: Mary Ann Liebert; 2009. p. 51

[68] Jessen AB, Toubro S, Astrup A. Effect of chewing gum containing nicotine and caffeine on energy expenditure and substrate utilization in men. American Journal of Clinical Nutrition. 2003;**77**:1442–1447

[69] Bisht S, Sisodia SS. *Coffea arabica*: A wonder gift to medical science. Journal of Natural Pharmaceuticals. 2010;1:58–66

[70] Belguidoum K, Guebailia HA, Boulmokh Y, Houache O. HPLC coupled to UV–vis detection for quantitative determination of phenolic compounds and caffeine in different brands of coffee in the Algerian market. Journal of the Taiwan Institute of Chemical Engineers. 2014;**45**:1314–1320

[71] Wagemaker TAL, Carvalho CRL, Maia NB, Baggio SR, Filho OG. Sun protection factor, content and composition of lipid fraction of green coffee beans. Industrial Crops and Products. 2011;**33**:469–473

[72] Lima AR, Pereira RGFA, Abrahão SA, Zangeronimo MG, Paula FBA, Duarte SMS. Effect of decaffeination of green and roasted coffees on the in vivo antioxidant activity and prevention of liver injury in rats. Brazilian Journal of Pharmacognosy. 2013;**23**(3):506–512

[73] Ochiai R, Sugiura Y, Shioya Y, Otsuka K, Katsuragi Y, Hashiguchi T. Coffee polyphenols improve peripheral endothelial function after glucose loading in healthy male adults. Nutrition Research. 2014;**34**:155–159

[74] Chandra S, Chatterjee P, Dey P, Bhattacharya S. Evaluation of in vitro anti-inflammatory activity of coffee against the denaturation of protein. Asian Pacific Journal of Tropical Biomedicine. 2012;**2**(1):178–180

[75] Burdan F. Caffeine in Coffee. Coffee in Health and Disease Prevention. London: Elsevier; 2015. pp. 201–207

[76] Armanian AM, Iranpour R, Faghihian E, Salehimehr N. Caffeine administration to prevent apnea in very premature infants. Pediatrics and Neonatology. 2016;**57**(5):408–412

Chapter 3

How Much Caffeine in Coffee Cup? Effects of Processing Operations, Extraction Methods and Variables

Carla Severini, Antonio Derossi, Ilde Ricci,
Anna Giuseppina Fiore and Rossella Caporizzi

Additional information is available at the end of the chapter

http://dx.doi.org/10.5772/intechopen.69002

Abstract

About 80–90% of the adults are regular consumers of coffee brews. Its consumption has positive effect on energy expenditure, power of muscle, while over consumption has negative effects widely debated. Across geographical areas, coffee brews may notably change when preparing Espresso, American, French, Turkish, etc. This chapter reviewed the phases able to affect the amount of caffeine in cup. Three most important areas will be addressed: (1) coffee varieties and environment; (2) coffee processing operations; (3) brewing methods extraction variables. What arises from the state of art is that, although there is a significant agreement on the effect of each critical variable on caffeine extraction, there is also a great difficulty to precisely know how much caffeine is in a coffee cup, although this is the most important information for the consumers. The number of affecting variables is very high, and some of them are inversely related with caffeine content (brewing time and brew volume), while others exhibit a direct relationship (grinding level, dose, and tamping). Finally, some variables under the control of barista rarely are accurately reproduced during brewing. For instance, it was found that the caffeine content in a Starbuck's coffee cup during different days varied significantly.

Keywords: caffeine, coffee, extraction, processing conditions, effect of variables

1. Introduction

The most studied component of the coffee is certainly caffeine (1,3,7-trimethylxanthine). It is present in the form of salt of chlorogenic acid and, in the roasted coffee, in free form. The caffeine amount present in raw coffee can significantly vary, depending on many factors, among which the most important are the origin and cultivar, Arabica or Canephora (var. Robusta).

On average, the raw Arabica shows a caffeine content ranging from 0.9 to 1.5% (dry weight), while the Robusta contains about twice as much between 1.2 and 2.4% [1–5].

But it is far from being considered a bad news. The World Health Organization (WHO) considers the coffee a "non-nutritive dietary component" because of its 2 calories per cup of bitter coffee. In fact, the numerous compounds formed during the roasting process come primarily from Maillard Reaction, and are considered as fiber. Like caffeine, they are hydrosoluble and can be easily disposed by the kidneys. From this point of view, the coffee and caffeine seem to be "neutral" component of human diet. Nevertheless, this is not true.

The positive effects of caffeine on the human organism are now widely known, with particular reference to the improvement of cognitive skills, as a stimulant of attention and concentration. From this point of view, coffee can therefore be considered a "functional" product according to the European Parliament and Council Regulation No. 1924/2006 of 2006 December 20 on "Nutrition and health claims made on foods" [6], it responds to the claims of type A which is related to the "improvement of a biological function related to specific physiological, psychological, and biological activities, beyond their established role in growth, development, and other normal functions".

However, in the past, caffeine was often demonized as responsible for diseases. Today, in the lights of the numerous studies conducted worldwide, it can be stated that caffeine is neither responsible for any disease related to cancer, development of cardiovascular diseases [7], nor related with problems that may arise during pregnancy, such as a shorter gestation or reduced birth weight (from a study of 12,208 women) [8]. Also in breastfeeding, the nurse can continue to drink coffee. It has been observed, in fact, that in the milk of women who drink coffee, caffeine reaches its maximum rate after about 1 h. Its concentration depends on the fat content of milk, and the infant absorbs only 0.06 - 1.5% of caffeine. So, there is no justification to prohibit nurses from a moderate coffee consumption [8].

On the contrary, a study carried out in Bhabha Atomic Research Centre in Bombay demonstrates that caffeine is able to contrast and prevent oxidative damage of human organism cell membranes, caused by free radicals, and shows an antioxidant capacity similar to the glutathione (antioxidant naturally present in the human intracellular fluid) and greater than the vitamin C [9].

From this point of view, the coffee, simply by the presence of caffeine, can also be considered a "functional" product that responds to the claims of type B related to the "reduction of disease risk that relate to consumption of a food or a food component that might help to reduce the risk of a given disease or medical condition because of specific nutrients or non-nutrients contained in it".

These topics about coffee, caffeine, and human health will be deeply discussed in Section 2 of this chapter.

Concerning the features of coffee beverage, it is known that its quality depends on a number of variables, so that starting from the same raw material we could obtain a coffee with completely different characteristics, in terms of pH, flavor, "body", cream, caffeine, phenols, Maillard Reaction Products (MRPs), etc.

For this reason, the chapter will consider all the principal aspects, which affect the quality of coffee "in cup".

Section 3 deals with the properties of raw material, the green coffee, and how its variability could affect the final quality of coffee brew.

Section 4 is aimed to deepen the central technological process that coffee undergoes during roasting. In this case, the different process conditions, applied in different countries and cultures, may lead to a range of possible chemical compositions (i.e. phenol content and MRP content) and sensory characteristics of coffee "in cup" (acidity).

In Section 5, the author investigates a particular aspect, which has been neglected so far, however being one of the most important for the quality of coffee beverage. The grinding process, which could dramatically affect all the features of beverage, such as volume, total solid content, caffeine content, pH, and flavor in general.

Sections 6 and 7 are totally dedicated to the different types of extraction and their fallout on caffeine content and other characteristics of coffee "in cup". Obviously, the single-service size systems are also described, even considering the more recent results of our research.

2. Coffee consumption and its debate on health

Coffee is an extremely popular beverage, which has become the second most valuable commodity after oil [10]. Annually, 120 million of coffee bags are consumed in the world, corresponding to over 7 million of tons [11]. Coffee consumption is a regular part of daily life worldwide [12], in fact in the European Community, as well as in the United States, the average consumption of coffee per capita is of 5.1 kg/year [13]. Americans consume more than 400 million of coffee cups daily, making this beverage the major source of caffeine in the adult diet [14].

The coffee is a complex mixture of thousands of chemicals. It contains, besides caffeine, more than 1,000 chemical compounds responsible of its flavor and aroma, carbohydrates, lipids, nitrogenous compounds, vitamins, minerals, alkaloids, and phenolic compounds [15, 16]. Anyway, among these, the caffeine is that on which is focused the majority of debates regarding the coffee consumption and its effects on health. Caffeine content in coffee is highly variable depending on a huge number of factors, such as variations in environmental and climatic conditions, features of raw materials, agricultural practices, post-harvest techniques, duration and conditions of storage, roasting degree, roasting process, type of commercial coffee, grinding, and brewing methods [16]. Caffeine is an alkaloid that is found in more than 60 plants which has a protective effect against insects [17]. The world's primary sources of dietary caffeine are roasted coffee beans and tea leaves. Caffeine is the most widely consumed psychoactive substance throughout the world, and it has been used for thousands of years [18].

Other common sources of caffeine are the kola nut, cacao bean, yerba mate, and guarana berries [19]. It has been estimated that 80–90% of adults are regular consumers of caffeine-containing brews, such as tea, coffee, cocoa, cola, and energy drinks [20]. A study on the caffeine intakes of the US population (considering a total of 37,602 consumers) showed that 85% of people consumed at least one caffeinated beverage per day. Caffeine intake in adults increases by age with the highest consumption for people of 50–64 years old (226 mg/day). Adult men consumed more total caffeine from beverages than adult women, as confirmed by Frary et al. [21]. In particular, the most frequently consumed beverages containing caffeine are coffee (71%), soft drinks (16%), and tea (12%) [19]. Although each of these has strong economic, social, and cultural impact, coffee brew remains the most important both economically and socially. In fact, coffee brew significantly contributes to the overall caffeine consumption of the adult populations [22].

In 2012, FDA [23] stated that for healthy adults, a caffeine intake up to 400 mg/day is not associated with adverse effects. Obviously for children, different and specific recommendations exist. Health Canada issued recommendations in 2009 specified the caffeine intake at 45–85 mg/day as healthy levels for children aged 6–12 years and 100–175 mg/day for adolescents of ages >12 years [24]. Health Canada recommended, for pregnant women, a daily dose of caffeine lower than 300 mg, while UK Food Standard Agency restricted this amount below 200 mg/day [12]. Brent et al. [25] and Peck et al. [26] do not support adverse effects for this caffeine consumption on reproductive health or pregnancy outcomes. Furthermore, another study [8] found that the caffeine consumption is not related with problems that may arise during pregnancy, such as a shorter gestation or reduced birth weight (from a study of 12,208 women).

Caffeine is rapidly absorbed in the stomach and small intestine, and it is distributed to all tissues, including the brain. Once caffeine is absorbed, it exhibits numerous and well-studied physiological effects.

However, apart from caffeine, coffee brews are also rich in other bioactive substances with a wide range physiological effects [27]. The list comprises of many phytochemicals, such as phenols, lactones, niacin, trigonelline, melanoidins, choline, etc.

An understanding of the physiological effects of coffee beverage is limited by the wide array of components included in the extracted product and by the numerous effects of each of these compounds.

However, it can be stated that the majority of the research carried out on the physiological properties of coffee has concerned the caffeine, which principally has stimulatory effects, including enhanced perception, reduced fatigue, enhanced memory consolidation, improved mental alertness, and reduced sleep duration [28]. A moderate consumption of caffeine has shown to increase strength and power of muscle, as well as energy expenditure. In fact, the consumption of 300 mg caffeine per day increases energy expenditure by approximately 79 kcal/day [29]. Moreover, it enhanced lipid oxidation and lipolytic and thermogenic activities [19].

On the other hand, an over consumption of caffeine might have negative effects, such as ringing in the ears, mood diarrhea, delirium, muscle tension, gastric acid secretion, etc. [30]. Excess caffeine intake is also involved in a state of excitement, anxiety, tachycardia, headache, palpitations, insomnia, nervousness, and tremor [31].

Wide differences in the dose-response of caffeine among individuals were observed as a result of genetic variation of susceptibility [32].

Furthermore, experimental and clinical evidences confirm tolerance from caffeine, which produce a reduction in the response as a consequence of previous exposure; consequently, the observed effects after a series of repetitive caffeine dosage may be very different from those highlighted after the first intake.

Apart from the well-known physiological properties of caffeine, more recent investigations indicated potential healthy effects of coffee, which are to a certain extent correlated with caffeine [31].

Some epidemiological studies suggested that coffee beverage is inversely associated with risk of various diseases [16, 33, 34].

Most of the more recent studies reported a relationship between a significant risk reduction of 30–60% in the development of type 2 diabetes and coffee consumption [34]. In particular, some studies reported a significant dose-dependent reduction in the risk of developing type 2 diabetes with a long-term coffee consumption [35, 36]. Moreover, this positive effect was observed both for caffeinated and decaffeinated coffee [37]; thus, it is possible to ascribe these effects to other phytochemicals.

Some studies reported controversial effects on the post-prandial glucose peak [38] as affected by coffee consumption. As reported from Greenberg et al. [39], part of these effects might be attributable to caffeine.

Furthermore, coffee intake has shown to reduce the liver damage in people at risk for liver diseases, such as hepatic injury, cirrhosis, and hepatocellular carcinoma [40, 41]. It was suggested that the coffee may preserve hepatocytes from damage, regardless of whether the aggressive agent is a virus, alcohol, drugs, or others [42]; however, the mechanisms associated with the protective effect of coffee on the liver are still unclear.

Coffee consumption is also inversely associated with the risk of Parkinson's disease in men and women, who have never used postmenopausal estrogen [43]. A meta-analysis found a risk reduction of 49% by consuming three additional coffee cups per day, whereas no effects were found for the cohort study that included only women [44, 45]. The well-reported protective effect of coffee on Parkinson's disease could be ascribed to its caffeine content, which acts to the dopaminergic system [31]. However, the mechanisms involved were not fully understood. Still about the risk of neuro-degenerative diseases, coffee drinkers have a lower risk of Alzheimer's disease respect to people who do not drink coffee [46], even if this outcome is under debate.

Some experimental studies asserted that cognitive deterioration of Alzheimer's disease in the central nervous system may be prevented by caffeine and/or chlorogenic acid [47, 48]. In addition, Gelber et al. [49] pointed out neither coffee and caffeine intake could be associated with any form of cognitive deterioration.

Regarding cancer, coffee consumption is inversely correlated with the risk of liver and colon-rectum cancers, even if the mechanisms involved are not clear yet [27]. Moreover, two meta-analysis concluded that there is a clear dose-dependent inverse association between hepatocellular cancer and the increase in coffee consumption, suggesting that by raising the intake of coffee, the possibility of developing hepatocellular carcinoma may be reduced [50, 51]. Also, a strong protective association has been found between coffee consumption and the reduction of endometrial cancer [52], while coffee intake might be weakly associated with breast cancer risk [53]. To highlight the protective effect of coffee extracts, it is worth noting that any association between these diseases and decaffeinated coffee was not observed. Under these considerations, it is possible that caffeine might be responsible for the protective role [31].

Only few studies have linked coffee consumption with an increased risk in developing cardiovascular (CV) disease. However, this risk is related to the ingestion of the diterpenes cafestol and kahweol, which have been shown to increase serum total and LDL cholesterol [54]. These compounds are mainly found in high amounts in boiled and unfiltered coffee. Besides these diterpenes, caffeine might exert negative effects on CV health too, by increasing heart rate and blood pressure [31, 55]. In a paper just published by a group of researchers from California [56] the effect of a diet rich in caffeine (coffee, tea, and cocoa) on the electrocardiographic profile of 1,388 study participants was tested. The subjects were followed up with clinical analysis and annual or semi-annual visits for 10 years and contacted every 6 months after this period. From results, there is no evidence (95% confidence) that frequent consumption of products containing caffeine is associated with heart problems. Patients with a history of heart problems showed no induction or cardiac arrhythmia aggravation within 1 h of taking 2 or 3 cups of coffee (275 mg caffeine). Moreover, one study involving about 3,000 patients hospitalized for cardiac arrhythmia showed an inverse relationship between consumption of coffee and caffeine and frequency of hospitalizations for arrhythmia, suggesting that it is highly unlikely that caffeine intake increases the risk of arrhythmia [57].

Coughlin and Nehlig [7] conducted a large study, which collects all the data made available by the worldwide research over the last 30 years, considering the balance of risks and benefits of coffee consumption as a whole. There is a plethora of potentially carcinogenic compounds (tested at high doses in animals) in coffee, but considered within the whole food ("whole food approach"), they produce a protective effect against many forms of cancer (lung, bladder, colon-rectal, endometrial, liver, prostate, leukemia, mouth, and throat). It is what the authors called "coffee paradox".

Therefore, the coffee beverage is now an important item in the lives of billions of people which is traditionally used to complement meals, as well as for hedonistic and psychostimulant purposes. Epidemiological data support the view that habitual coffee consumption has several health benefits because of its content of bioactive compounds and caffeine, which can exert physiological and healthy effects. Caffeine intakes up to 400 mg/day do not give rise to safety concerns for healthy adults in the general populations.

3. Coffee species, origin and blending

Coffee's most studied component, caffeine, varies substantially as a function of coffee plant species [58]. Green coffee beans are used by the International Standard (ISO 3509-1989) to define *"a commercial term designating the dried seeds of the coffee plant"* [2]. Coffee beans are produced from the cotyledons of seeds belonging to the genus *Coffea*, which includes approximately 70 species. Some of these are of small-scale, they are cultivated in some African countries, but the resultant beverages are generally of low quality and most of the beans are not exported [59]. Three coffee species are mostly commercialized: *Coffea arabica, Coffea canephora Pierre*, and *Coffea liberica Bull* worldwide known as Arabica, Robusta, and Liberica or Liberian coffee, respectively. However, only the first two have a commercial importance; in particular, *C. arabica* provides for 60% of world production, while the remaining 40% are from *C. canephora* var. Robusta (**Figure 1**) [60]. These two species display differences, deriving from optimal climate of growing, physical aspects, chemical composition, and quality of the beverages. Generally, coffee extract prepared by *C. arabica* is more appreciated than Robusta because of its superior quality in terms of aroma and, therefore, it reaches higher prices in the international market [61]. On the other hand, the Robusta coffee, characterized by a more bitter and persistent taste, shows a high amount of antioxidants and soluble solids [3]. However, green beans are especially featured by their content in caffeine, trigonelline, and chlorogenic acids. The two main species exhibit differences in caffeine percentages ranged between 1.2-2.4% for Robusta and 0.9–1.5% for Arabica [1–5]. Caffeine is formed in immature coffee fruits, and it gradually accumulates all along seed development [62]. The lower content in caffeine for *C. Arabica* is explained by a lower expression of some genes (CaXMT1, CaMXMT1, and CaDXMT2) respect to *C. canephora* [63]; these genes were positively correlated with the caffeine accumulation in coffee beans. Likewise, geographical origins may have influence on caffeine accumulation and its final concentration. Babova et al. [5] reported that Arabica coffee from Brazil contains more caffeine than the same species growth in Ethiopia and Kenya; similarly, the Robusta coffee from Uganda shows more caffeine than the same species coming from Vietnam. Furthermore, Cheng et al. [63] deeply reviewed the metabolism of the most important components in coffee as affected by genotype (G) and environment (E), showing as both affect seed development and the final concentration of metabolites in coffee beans,

Figure 1. Green beans of *C. arabica* and *C. canephora* var. Robusta.

especially caffeine content. Specifically, the authors highlighted as G and E, as well as their interaction (G×E), may affect the overall quality of coffee; similar results were found in a recent study, in which 20 samples of *C. arabica* and *C. canephora* were investigated [5]. This study highlighted a clear separation among *C. arabica* accessions based on their geographical origins, with Ethiopia and Mexico's accessions which exhibit the lowest content of caffeine [5]. Despite the high differences between the two most important species of coffee beans, in terms of caffeine content and geographical origins, there are variations within the same species and across the different cultivars [5].

Other than genotype and geographical origins, other environmental factors may affect caffeine accumulation. For example, light exposure is required for caffeine synthesis, although its optimal level is very low [63]. Indeed, some researchers demonstrated as increased level of shade improves caffeine content in *C. arabica* [64, 65], while seedlings of Robusta coffee completely grown in darkness showed a remarkable decrease in caffeine content [66]. Furthermore, among environmental conditions, also the high altitude was positively related to caffeine content [67].

Fox et al. [68] studied the variations in caffeine concentration for 25 single beans from 5 selected coffees. They found a positive relationship between the weight of beans and caffeine content, but a very low determination coefficient, r^2, of 0.31 was calculated. This proved that selecting the beans for weight would not ensure an increase in caffeine concentration.

However, apart from geographic origins, rarely the coffee used to prepare a beverage consists in a unique species; the blending is a technique used to improve the overall aroma, body, and flavor of coffee, with the main aim to obtain a coffee having excellent sensorial properties on the final roasted product. Blending may be done before or after roasting, even though, traditionally, the retailer and the roasters perform it before the roasting, by combining green beans with similar characteristics, to obtain same physicochemical changes during the thermal process. To date, many popular blends are available on the market which may show notable changes in caffeine content based on origin, species, cultivar, and ratios used in blending. Generally, commercial blends available in the market present a great variability, mainly depending on the species used, even if other factors may influence caffeine content; for example, roasting degree, grinding level of coffee powder, etc. which will be well presented in the next section.

4. Changes in caffeine content as affected by roasting

Roasting is one of most important step in coffee processing because of the marked chemical, physical, structural, and sensorial changes that confer the worldwide appreciated properties. During this process, coffee beans are exposed to high temperature for a time length that can greatly vary according to the type of roaster, geographical origin, variety, coffee bean characteristics, and the desired sensorial properties. Coffee roasting is a process carried out in different ways throughout the world (**Figure 2**) [69].

http://dx.doi.org/10.5772/intechopen.69002

Figure 2. Coffee roast levels (adapted from Ref. [69]).

In terms of structure, the beans increase their volume becoming up to almost a double of the original. Moreover, the beans lose weight in a range of the 15–25% as well as a continuous change in color is commonly observed (**Figure 3**) [70]. The modifications involved during the roasting are the result of hundreds of chemical reactions and thermal decompositions occurring on thousands of chemical compounds. Four regions of decomposition of the green coffee beans have been reported: (1) dehydration; (2) hydrolysis; (3) desmolysis; (4) catalysis [71]. The decomposition begins at 100°C by a significant endothermic reaction that is followed by a drop in temperature (**Figure 3**) [72, 73]. Among the reactions occurring during this endothermic step, the major contribution seems to be given by the phenols. Considering caffeine, in spite of its high sublimation point (178°C), a reduction is observed by evaporation because it is dragged by the water vapor [73]. This phenomenon is also allowed by the increase of the caffeine solubility in water as a function of temperature.

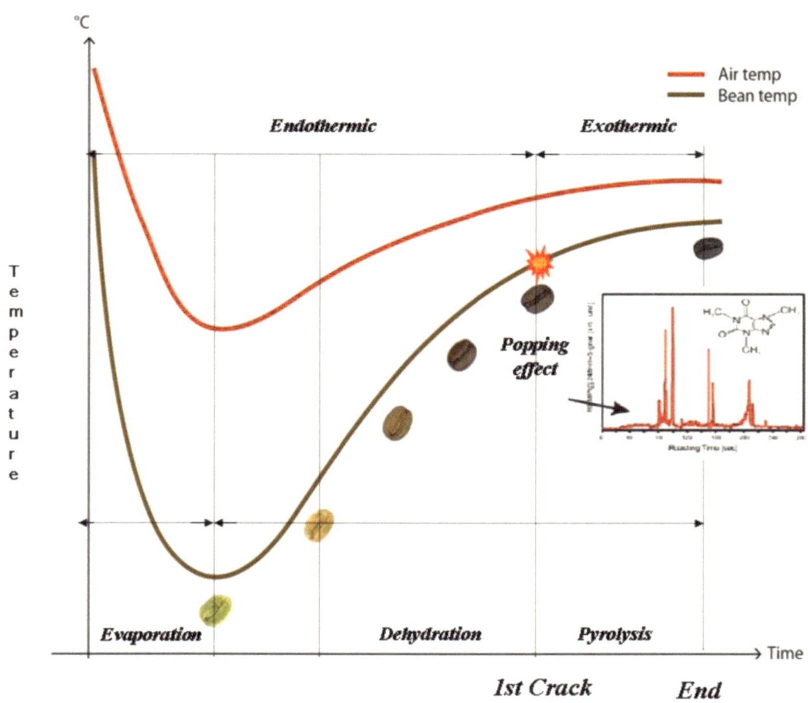

Figure 3. Roast profile analysis (adapted from Refs. [70, 80]).

In general, the roasting causes a reduction in caffeine content of 30% (from 0.89% ± 0.02 of green beans to 0.6% ± 0.03 for roasted Arabica beans) [74]. Farah [3] confirmed that even though caffeine is not involved in chemical reactions, being stable upon roasting, a small fraction may be lost by sublimation. Analyzing the evolution of gas composition during roasting, it was observed that in the same temperature region (100–245°C), an increase of nitrogen-containing heterocyclic compounds, such as indole and caffeine occurs.

The stability of caffeine, during roasting process, was also reported for the roasted coffee oil (an important by-product with aromatic properties of the Brazilian soluble coffee industries) obtained by mechanical pressing of beans before the extraction of soluble coffee. During mechanical *via* expelling extraction (high pressure and high temperature), a large amount of caffeine is incorporated into the roasted coffee oil because it is not thermally degraded [75].

Due to the temperature of sublimation (178°C) [76, 77], it would be expected that the loss of caffeine would occur to a higher extent when this temperature is reached. Macrae [78] reported that these phenomena could be related with porosity and the internal pressure created into the beans that may cause some difficulties for the sublimation of caffeine. Nevertheless, in a model system, where caffeine is probably free of chemical and physical linkages, a similar gradual decrease of its content occurs [79].

Moreover, important microstructural changes occurring during roasting can drive an additional loss in caffeine. The high temperature reached during roasting causes bursts accompanied by popping sounds [80]. During popping phenomena, caffeine is easily detectable in the roasting gas, because it is emitted during seed fracturing (**Figure 3**). Popping is a consequence of the accumulation of inorganic gases formed into the closed pores of beans, during the pyrolysis of several compounds. When the pressure reaches a critical limit, the seeds crack and the entrapped gases are abruptly released. Under these conditions, darker roasting degrees could present less caffeine amount.

However, the roasting variables may be classified as intrinsic and extrinsic process. The first class includes all that can be controlled and changed depending on the desired degree of roasting (methods of roasting, time and temperature profiles, and coffee's load), while the latter depends on the features of the green beans (variety, species, origin, and quality) and its pre-processing (batch-to-batch differences in the coffee beans, semi-dry or wet post-harvesting method, and humidity).

Among the extrinsic variables, Crozier et al. [12] reported that caffeine content depends on preliminary processing to which beans are subjected. For example, both the washed and unwashed Arabica beans submitted to different time/temperature profiles, such as high temperature for short time (H-S) and low temperature for a longer time (L-L) led to a reduction in caffeine of 80% (in comparison with green beans) in the corresponding coffee brews prepared by adding 5 g of ground beans in 100 mL of boiling water for 5 min than unroasted samples. However, the brew obtained from washed Arabica roasted beans retained the 20.6 and 19.6% for H-S and L-L, respectively, while a better retention of caffeine was observed when using unwashed Arabica beans with values of 19.2 and 18.6% for H-S and L-L, respectively. Coffee bean's humidity markedly affects the time of roasting: as well as the temperature of the beans at the end of roasting.

Taking into account roasting techniques, coffee beans are traditionally roasted in batch, working hundreds of kilograms, or in continuous systems. The heat can be transferred to the beans by conduction at direct contact with hot metal surfaces, by free or forced convection due to a streaming media (hot air), or by radiation [81]. Moreover, non-conventional microwave roasting or combined techniques were also studied [82, 83]. The authors reported that the application of microwave roasting determined a lower loss in caffeine (10.38%, from 2.12 to 1.90 g/100 g) rather than conventional roasting (14.15%, from 2.12 to 1.90 g/100 g). However, combined methods (convective and microwave) enabled to obtain a further preservation in caffeine content exhibiting a total loss of 8%.

The microwaves operate directly in the core of the beans, so that the process of roasting is intensified throughout the whole interior of the bean. This leads to a very intensive heating from the core to the surface of beans. The application of combined methods resulted in the increasing of heating and chemical reactions, a reduction of roasting time, while the ultimate temperature of coffee is lower than the values measured by traditional convective heating. The same changes in caffeine content were observed by headspace analysis of corresponding ground of green and roasted coffee beans (**Table 1**) [82, 83]. Because of these reasons, the microwave roasting method was found to be the most advantageous for caffeine retention.

Another key factor for the process is, of course, roasting time. The quantity of heat transferred to the beans is the result of temperature and roasting time [84]. According to the widespread opinion, the degree of roast in the product is correlated to the final roasting temperature [84, 85]. In general, temperature must exceed 190°C to provide a sufficiently reactive roast environment; therefore, the residence time and the process temperature should be precisely measured to describe the overall thermal behavior. For example, **Table 2** reports the changes in caffeine for Arabica and Robusta coffee beans during two roasting experiments [79]. The first trial was performed at constant roasting time of 15 min by increasing temperature, while the second one was performed by increasing roasting time at fixed temperature of 240°C. At constant roasting time, caffeine content decreases of 11.3% (from 1.24 to 1.10 g/100 g d.w.) and 7.7% (2.08 and 1.92 g/100 g d.w.) in Arabica and Robusta coffee beans, respectively. Roasting temperatures until 220°C did not caused any loss in caffeine content in Arabica coffee beans, while a slight decrease of 4.3% (from 2.08 to 1.99 g/100 g d.w.) in Robusta coffee occurred

Roasting method (medium roast degree 9.5% of solids substances)	Roasting time (min)	Beans temperature (°C)	Caffeine in beans (g/100 g d.w.)	Caffeine in headspace surface area of GC peak (%)
Unroasted bean	–	–	2.12 ± 0.03	2.01
Convective	9.75 ± 0.21	238 ± 3	1.82 ± 0.02	0.23
Microwave	11.08 ± 0.17	207 ± 2	1.90 ± 0.05	0.30
Convective-microwave	5.33 ± 0.12	195 ± 0.08	1.95 ± 0.08	0.37

Table 1. Caffeine content in green and roasted beans (g/100 g dry weight) and in corresponding headspace after roasting process with different methods (adapted from Refs. [82, 83]).

Experiment 1			Experiment 2		
Roasting temperature (°C)	Caffeine (g/100 g d.w.)		Roasting time (min)	Caffeine (g/100 g d.w.)	
	Arabica	Robusta		Arabica	Robusta
Green	1.24	2.08	Green	1.24	2.08
140	1.44	2.21	5	1.14	2.04
160	1.52	2.17	8	1.14	2.04
180	1.36	1.98	12	1.05	1.98
200	1.39	1.99	15	0.99	1.93
220	1.29	1.99	20	0.93	1.91
240	1.10	1.92			

Table 2. Changes in caffeine content during roasting at fixed time of 15 min (experiment 1) and at fixed temperature of 240°C (experiment 2) (adapted from Ref. [79]).

probably due to its higher water content. At constant temperature of roasting (240°C), caffeine content decreases of 20% (from 1.24 to 0.93 g/100 g d.w.) and 7.21% (2.08 to 1.91 g/100 g d.w.) in Arabica and Robusta coffee, respectively, after 15 min of process. These data state that the temperature, rather than time, is the main factor affecting caffeine loss during roasting. Therefore, the evolution of caffeine content from raw or green to the roasted beans depends to the chemical and physical changes that occur during process. Although it not degrades, caffeine content could be reduced in two phase of roasting, during dehydration in which caffeine is dragged by water vapor and during the first crack of the beans, with other volatile compounds, as well as when its sublimation temperature is reached. However, this slight reduction is observed in the final step of roasting, determining a less caffeine concentration in the dark roasted beans.

5. Effects of grinding on caffeine extraction

The grinding is a crucial step for coffee brew preparation. In roasted whole beans, the volatiles and the chemical compounds are entrapped in cells and they barely can dissolve in hot water. After grinding, the beans are reduced to small particles having micro- and meso-scale dimension (from few micrometers to ~1,000 μm) from which volatiles may be released and chemical compounds are easily dissolved in hot water, giving the worldwide appreciated aroma [86]. Consistently, from coffee powder about the 60% of aroma is lost during the first 15 min after grinding. For this reason, coffee brews should be rapidly prepared, with the aim to keep its aroma as much as possible. Moreover, in terms of chemical compounds dissolved in coffee brew, the grinding process is one of the most important critical control points for extraction phenomena. Moroney et al. [87] stated that "*particle size of coffee ground is vitally important in coffee extraction in that it affects both the fluid flow through the grind and the grind's extraction kinetics*". Commonly, ground coffee is classified in four groups, such as coarse, medium, fine, or very fine. However, across different countries various particle size distributions may be indicated with the same name, as in the case of Europe and USA where the coarse coffee ground has an average size of 850 and 1,130 μm, respectively, likewise the fine ground coffee, which shows an average size of 430 and 800 μm, respectively [2]. The percolation of water inside the voids (capillaries) in coffee cake, the wettability of each coffee particle, and the diffusion of chemicals from coffee particles to hot water are the main phenomena controlling the amount of chemical compounds released in coffee beverage [88]. When coarse particles are used, the percolation rate is high, due to the greater porosity fraction of coffee cakes and the dimension of its capillaries. This condition leads to an overall decrease in extraction of chemicals. Moreover, diffusion process is reduced due to the decrease in surface contact area between particles and hot water. On the other hand, fine or very fine coffee ground may create a coffee cake very close to its percolation threshold. In this case, the extraction time significantly increase, and a different extraction may occur. A proper equilibrium between percolation, diffusion, and wettability of coffee particles drives the type and the amount of chemicals in coffee hence its quality in cup. Therefore, as a rule, the grinding must be adjusted on the basis of the sensorial and chemical properties desired in coffee brew (i.e. the type of coffee brew). French press coffee, for which the infusion of coffee ground in hot water takes several minutes, needs coarse particles with the aim to get slower diffusion avoiding the extraction of bitter compounds. When preparing espresso coffee with automatic machines, working under pressure, extraction time is reduced to 25–30 s, and finer particles are needed to increase extraction rate of chemicals and volatiles. For French press coffee, about 100–300 particles are usually obtained from each coffee bean, while 3,500 and 15,000–35,000 particles are obtained for preparing Espresso and Turkish coffee, respectively [89]. **Figure 4** schematically depicts the overall particle size distribution for the most common coffee preparation. However, a bimodal particle size distribution is generally preferred being

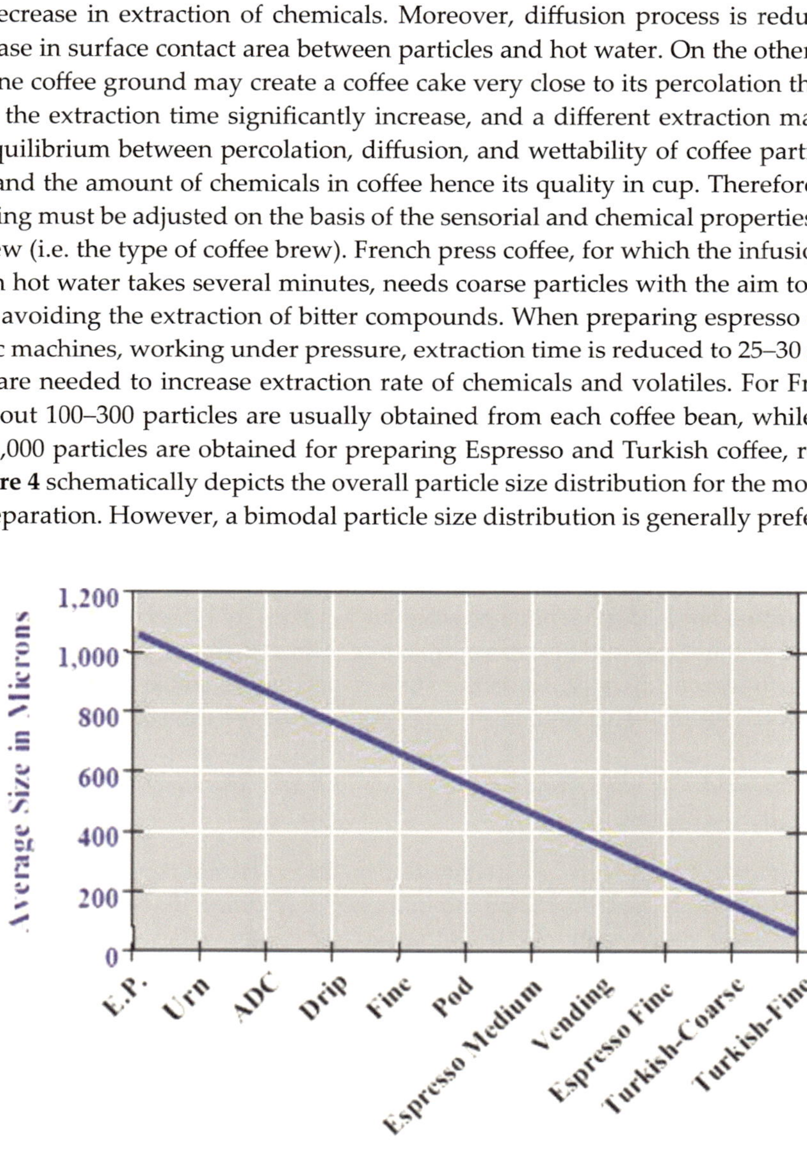

Figure 4. Average particle size of coffee ground for different preparation methods [89].

it able to keep a good equilibrium between wettability, percolation, and diffusion phenomena [88, 90]. Petracco [91] reported that a bimodal distribution of coffee particles is the starting point to obtain a good espresso coffee. The result of grinding operation is affected by several variables, such as the mechanical properties of coffee beans, the moisture content of roasted beans, the type of grinders (blade grinder, conical, or flat burrs grinder). Also, the grinding affects the stability of coffee powder during storage being strictly related with the agglomeration phenomena and aroma retention [88, 92].

The grinding uniformity, that is, how large is the particle size distribution, of crucial importance. If it is poor, an extraction time ideal for the smallest particles will be incorrect for the larger ones, thus leading to a tea-like taste [89]. The impact of particles size of ground coffee on the quality of the brew was widely studied by several authors [90, 93–96]. However, the effect of grinding on the caffeine extraction and its amount in cup have not been studied yet in details.

Spiro and Selwood [97] explored the effects of particle size on the kinetic of caffeine infusion. By separating the coffee ground in sub-groups of particles having different size, the authors estimated the rate constants for extraction caffeine by infusion in water. They reported an increase from 0.207×10^{-3} to 22×10^{-3} s^{-1} for particles size of 1,700–2,400 and 152–211 μm, respectively. Of course, this is in accordance with the general decrease of coffee particle-water contact area. Bell et al. [93] studied the effect of grinding level on the caffeine content of coffee brew. Although the authors did not analyze the particle size distribution of coffee ground, they showed that by using 8 g of coffee ground for 355 mL of filtered brew, the finest powder yielded the higher caffeine content of 70 mg/177 mL, while when the coarse coffee ground was used, the caffeine content was of 50 mg/177 mL. Again, this was an effect of the greater surface contact area between the fine coffee ground and hot water, which favored the caffeine extraction. On the other hand, when the authors used 32 g of coffee powder for 1,420 mL of water any difference in caffeine content was not observed by using coarse-medium or coarse ground. In spite of the same ratio coffee ground/water, when the authors used more coffee ground, the grinding levels did not have effect. This was due to the longer extraction time of 10 min during which the caffeine was completely extracted, independently from the particles size. Instead, when 8 g of coffee ground was used, for a brewing time of 3 min, the effect of the grinding levels was statistically significant.

An interesting result was obtained for people who prefer to perform the grinding at home with commercial grinders. By home-grinding, no influence of grinding time on caffeine content of the brews was observed. As reported from the authors, the low efficacy of the home-grinder produced very large particle size distribution function being overlapped for 8 or 18 s of grinding time. The authors used the term "*less distinctive grinding patterns*" to explain that no statistical differences were observed increasing the grinding time of 10 s. Similar results were reported by Buchmann et al. [98], who studied the impact of grind size, water temperature, and coffee/water amount on trigonelline and caffeine in Espresso and American brew coffee. In accordance with above discussion, the authors showed the increase of caffeine content from coarse to fine particles. For Espresso coffee, values of ~25 mg/65 mL, ~62 mg/65 mL, and ~75 mg/65 mL were measured by using 7.5 g of coarse, medium, and fine coffee ground, respectively. Similarly, for Fresh Brew (American filtered coffee), values of ~45 mg/125 mL, ~65 mg/125 mL, and ~62 mg/125 mL were determined when 9 g of coarse, medium, and fine coffee ground were used. However, the authors did not report the particle size distribution of coffee grounds.

These observations enable to introduce the importance of the relationships among particle size, extraction time, and volume of the brew. Severini et al. [90, 95, 96] deeply studied how these variables affect the quality of espresso coffee. The authors analyzing the effect of using coarse, fine-coarse, and fine coffee ground (**Table 3**) [95] on the caffeine concentration collecting three brew fractions: the first 8 s (Ft1), from 9 to 16 s (Ft2), and from 17 to 24 s (Ft3). Without regard to the fraction time, the caffeine concentration exhibited the following order: fine > fine-coarse > coarse. For instance, values of 4.98, 4.35, and 2.41 mg/mL were measured for Ft1 samples [95]. It was highlighted that this increase was not only the result of the reduction of particles size, but also the consequence of a reduced brew volume for a less percolation rate that, in turn, was due to the lower porosity in coffee cakes.

Under this consideration, the authors modeled the caffeine extraction through coarse, fine-coarse, and fine coffee ground [90]. First, the authors proved that among grinding, doses, and tamping, the former was statistically the most important to explain the caffeine behavior during extraction. Nonetheless, when considering the total amount of caffeine in cup an opposite order was observed. For instance, the authors reported caffeine content of 75.60, 98.97, and 128.79 mg/cup for fine, coarse-fine, and coarse coffee ground after 14 s of extraction. The volumes of coffee brew, after 14 s, were 10, 22, and 50 mL for fine, fine-coarse, and coarse powder, respectively. By using coarse coffee ground, a greater percolation rate (i.e. the amount of water that flowed through coffee cake in the unit of time), due to the large pores available, increased the extraction of caffeine. In spite of the use of fine coffee powder gives a greater particles-water contact area, the lower percolation rate reduced the total amount of caffeine in cup. The authors stated that these results proved that the major contribution to the total caffeine content of espresso coffee in cup was given by the percolation rate, rather than the grinding level.

Particle size (μm)	**Grinding grade**		
	Fine	**Fine-coarse**	**Coarse**
>600	0.21 ± 0.11	0.68 ± 0.22	2.93 ± 0.53
$400 < X < 600$	5.69 ± 1.26	13.80 ± 0.82	33.87 ± 1.63
$250 < X < 400$	32.00 ± 4.89	47.15 ± 14.12	35.64 ± 1.88
$180 < X < 250$	52.60 ± 6.12	37.18 ± 14.10	26.42 ± 1.07
<180	9.52 ± 3.21	1.19 ± 1.02	1.15 ± 0.85

Table 3. Distribution (%) of particle size in each grinding grade of coffee powder (mean values ± standard deviation) (adapted from Ref. [95]).

6. Coffee preparation: methods

The consumer preferences in terms of the sensorial properties of coffee are affected by different factors, such as culture, lifestyle, social behaviors, habits, and economic aspects. Moreover, more recently, the attention of consumers is focused on the outcomes of coffee intake on health and well-being of specific components, such as caffeine and bioactive compounds. In this contest, brewing methods and the extraction conditions are essential to obtain the desired chemical, sensorial, and healthy properties of coffee in cup. A wide literature, across the last 20 years, is available but often the published data are difficult to compare due to the difference in coffee preparation conditions. On the other hand, all authors revealed that there is not *"the best coffee preparation method"*, but every extraction has its own peculiar characteristics [20, 99–104]. In the following sections, we present the most relevant data and discussion on the different brewing methods and their effect of coffee beverage quality.

6.1. Brewing methods: geographical and cultural aspects

Depending on geographical origins and cultural traditions, different brewing techniques are commonly used to make a coffee cup in the world. Among the most important and popular, a coffee cup may be prepared as Espresso, Turkish, American, Moka, Neapolitan, and French press coffee. However, as reported by Petracco [105] under a physical point of view, the coffee preparations may be classified in three main methodologies: (1) the "original Italian method" under high pressure (i.e. Espresso and Moka); (2) infusion by pouring hot water on ground coffee followed by a filtration (i.e. Drip filter, French press or Plunger, and Neapolitan); (3) decoction methods (i.e. Turkish, boiled, percolator, and vacuum). All these methods noticeably affect the type and the amount of chemical compounds extracted, including the caffeine content. Drip filter coffee is the most popular brewing method in the world. It is largely diffused in USA, while in north Europe, France, and Scandinavian region, the plunger or French press coffees are the most consumed. When considering the southern European countries, a greater variability in the coffee brew methods is observed. The Turkish coffee is consumed in the Middle East, North Africa, Balkans, Greece, Turkey, and various locations within Eastern Europe [106]. In Italy, Spain, and Portugal, coffee cups are generally prepared by using the Espresso method and Moka [100, 102, 103, 107, 108]. The instant coffee, also known as soluble coffee, initially consumed mainly in Great Britain and Japan, later has been spread all over the world [20, 109, 110]. Finally, in the last 10 years, the single-dose pod or capsule system has gained interest for the preparation of coffee at home or at work [104, 111–113].

6.2. Variables affecting caffeine content

Once the blending of coffee varieties, the roasting level likewise the grinding degree has been chosen, obtaining the desired roasted-ground coffee, several brewing methods may be used to prepare our coffee cup. However, for all of these, the theoretical principle consists in a solid-liquid extraction of all chemical compounds from roasted-ground coffee (soluble solid) to hot water (solvent) [114]. Considering the brew preparation at coffee shops, bar, or at home, several variables may modify the coffee quality in cup. The type of contact between water and coffee ground, the extraction time, the roasted-ground coffee/water mass ratio, the extract volume as well as water temperature, the vapor pressure in the case of Espresso coffee, filtration, and boiling process play important roles on the caffeine content of the beverage, as well as on functional and sensorial compounds [20, 102, 104, 105, 109, 115, 116]. First of all, the volume of the brew in cup is the variable exhibiting the wider variance mostly due to the personal appreciation. For instance, the coffee cup may vary from the "*Ristretto espresso coffee*" [117] of about 15–20 mL to the "*American filtered coffee*" of 125–400 mL [100]. This, of course, greatly affects the caffeine intake every day. Several studies indicated that the caffeine contents ranging from 2.4 to 4.5 mg/mL for Espresso (25 mL), from 0.4 to 1.4 mg/mL for American or filtered (200 mL), from 0.2 to 0.5 for French or Plunger (100 mL), from 0.7 to 5.4 mg/mL for Moka (30 mL), 1.6 mg/mL for Neapolitan (30 mL), and 1.94 mg/mL for Turkish (50 mL) coffee brews. This highlights that the volume of beverage per cup has a profound effect on the assumption of caffeine. A second variable, which significantly changes among the brewing method, is the powder/water (p/w) ratio. It was reported that 7 g/25 mL are commonly used to prepare an Italian espresso coffee, 12 g/200 mL are adopted for American or filtered, while 8 g/100 mL and 5 g/50 mL are used for French and Turkish coffee brew, respectively. Moreover, extraction time is subject to a huge variability. Taking into account the difference in coffee powder/water ratio, several authors highlighted that about 25 s are necessary to prepare an Espresso coffee and 5–7 min would be needed for American and French coffee brews [100–103, 107].

In the following sections, the most important brewing methods and the extraction conditions will be analyzed in detail, paying attention on the quality of the beverage, particularly concerning the caffeine content.

6.3. Espresso coffee

The Italian Espresso coffee (EC) is one of the most appreciated coffee brews, an intense aromatic beverage made for immediate consumption. EC may be defined as "*a brew obtained by percolation of hot water under pressure through compacted cake of roasted-ground coffee, where the energy of the water pressure is spent within the cake*" [84]. In general, an Espresso coffee (~25 mL) is prepared by ground roasted coffee beans (6.5 ± 1.5 g), by means of hot water (90 ± 5°C) under pressure (9 ± 2 bar) applied for a short extraction time (30 ± 5 s) to a compact roast and ground coffee cake by a percolation machine, to obtain a small cup of a concentrated foamy elixir [105]. However, important differences are commonly observed, such as the so-called *ristretto* with a volume of brew < of 20 mL and the *lungo* espresso coffee > of 30 mL, which are often consumed in Italy and in other countries [117]. Apart from the above overall definition, the in-cup quality of espresso coffee and, particularly caffeine content, is affected by several variables under control of the *barista* (espresso coffee professional bartender) as shown in **Figure 5**.

After the choice of coffee blend and roasting degree, the first step to produce the EC brew is the grinding of the roasted beans at the optimal level. As previously reported, grinding level of roasted coffee powder greatly affects the caffeine content in coffee brew. In general, until a certain level of finesse, a decrease of the particle size of the coffee ground implies an increase of the caffeine content due to the larger surface area. For an espresso cup prepared by fine ground

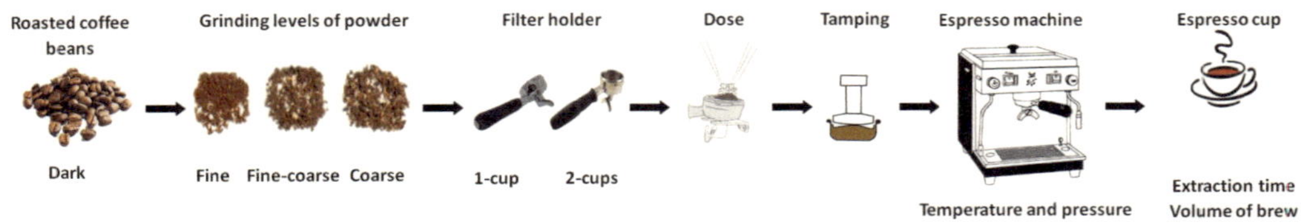

Figure 5. Espresso coffee: variables under control of the *barista*.

coffee, the caffeine content varies from 2.1 to 4.2 mg/mL, while using the coarse coffee powder, caffeine concentration was ranged from 0.5 to 3.2 mg/mL [90, 94, 95, 102]. Andueza et al. [94] reported that the caffeine in EC, from roasted coffee blend (20% Arabica – 80% Robusta), was of 3.80, 3.19, and 3.05 mg/mL for very fine, fine and coarse, ground coffee respectively. Severini et al. [90, 95, 96] confirmed that, maintaining constant the dose of coffee ground, the pressure on the upper surface of cake and the extraction time, the caffeine content in espresso was strictly correlated with the grinding level. In general, therefore, all authors agree with an increased caffeine extraction as finer is the ground coffee used for brewing. However, as reported below, several other variables may affect the caffeine concentration with some of these having a direct relationship, while others are inversely related. So, as explained in the following section, in some extraction conditions, the effect of grinding level could be also completely invalidated.

Among these, one of the most important is the dose of ground coffee, that is the amount of coffee powder used to prepare an espresso coffee cup. Romani et al. [118] reported that the dose of roasted coffee powder used to prepare a cup of espresso is found to be between 6 and 8 g, until a limit of 9 g. This variability has an important effect on the caffeine content of espresso coffee. Andueza et al. [116] highlighted that the caffeine content in EC cup is greatly affected by the quantity of coffee powder used. By preparing an EC of 40 mL, using 6.5, 7.5 and 8.5 g of ground coffee, the authors showed several differences in caffeine with values of 1.80, 1.88, and 2.21 mg/mL, respectively, when using doses of 100% Arabica coffee. Similarly, by using a blend of 20% Arabica – 80% Robusta, higher caffeine values of 3.01, 3.17, and 3.31 mg/mL were, respectively, obtained for the same doses.

However, the effect of ground coffee dose on caffeine concentration could be roughly analyzed without taking into account the corresponding amount of water used to prepare the brew (i.e. coffee powder/water mass ratio).

The analysis of this variable is correctly interpreted for brewing methods in which the amount of water is defined before coffee extraction, such as American coffee, Turkish, etc., but it would be wrongly analyzed for espresso coffee methods for which the variable ground coffee/water ratio is rather a ground coffee/brew volume ratio.

However, depending on the traditions, a cup of espresso coffee in Italy is of 20–25 mL, in Spain of 40–60 mL, and in Scotland about 30–50 mL. From these data, roasted coffee powder/water ratios were 7 g/20 mL, 9 g/60 mL, and 11 g/30 mL, and the caffeine contents resulted, in mean, of 5.4, 1.8, and 3.9 mg/mL, respectively [110].

Several studies highlighted that the increase in dose and/or grinding level, keeping constant the total volume of EC, determines an increase of caffeine concentration. Moreover, when the dose of ground coffee is higher, being the powder/water surface greater, the percolation pathway for hot water through the compact cake is more tortuous, increasing the brewing time and more aromatic and chemical compounds in coffee beverage [90, 116].

Furthermore, the pressure on the upper surface of the coffee cake (tamping) is a step of crucial importance for both the microstructural properties of the coffee cake and, therefore, on the pathway of water during the percolation [119]. Severini et al. [90] highlighted that differences may be in the chemical composition of espresso cup, including the caffeine content, applying, for 5 s, different pressures (0.75, 1.5 and 2.25 kg) on the coffee cake. On these basis, the tamping step which could be underestimated at bar could have a significant effect on the caffeine content of espresso coffee independently from the grinding level and the coffee powder/water ratio. Also, several studies on the espresso machine conditions (pressure and temperature) are available. Masella et al. [120] studied the effect of temperature and pressure of water on the quality of espresso coffee. They found that the combination between three temperatures (75, 80, and 85°C) and two machine pressures (15 and 20 atm) not influenced the caffeine content of the coffee samples, showing an average value of 2.25 mg/mL. These data well agree with Andueza et al. [121] who tested three water pressures (7, 9, and 11 atm) on caffeine content of espresso coffee showing a mean value of 2.04 mg/mL. On the other hand, Salamanca et al. [122] proved that by applying a gradient of temperature to prepare an espresso cup across different varieties of roasted coffee, an increase or decrease of some chemical compounds was highlighted, among these the caffeine.

The extraction time is also a crucial variable in terms of chemical extraction. Nicoli et al. [1] divided the volume of beverage in five fractions of 10 mL each during espresso coffee preparation. In the first fraction, the highest caffeine concentration with a value of 6.5 mg/mL was observed, while a value of 0.2 mg/mL was found in the last fraction. According to these data, Mora and Rodriguez [123] reported that are necessary only 10 s to extract the 60% of caffeine from roasted coffee powder (100% Arabica, 100% Robusta, and blend) when preparing an EC cup of 30 mL. Ludwig at el. [101], monitoring some chemical compounds during the extraction time, measured a caffeine concentration of 4.36 and 0.57 mg/mL in the first 0–8 s and 16–24 s, respectively. Severini et al. [90, 95, 96] proved that the extraction time highly affected the aromatic and chemical compounds of EC brew. Taking into account the caffeine content, it was shown as during the first 8 s of extraction, the caffeine concentration was comparable with the value measured in a cup of 25 mL. Therefore, all authors agree with progressive reduction of extracted caffeine as a function of brewing time, obviously caused by the reduction of remaining caffeine in ground coffee. In general, all papers confirmed that the majority of caffeine is extracted in the first phase of brewing. For example, considering a fine roasted coffee powder, the caffeine concentration was of 4.98 and 4.18 mg/mL after 8 s of extraction (Volume ~16 mL) and in the final EC cup (25 mL), respectively. Of course, the reduction of 0.8 mg/mL of caffeine in the final cup is due to the dilution effect since in the last seconds of extraction, only water falls in the cup, in practice. Using a coffee powder fine, fine-coarse (or medium), and coarse, it was reported that 22, 15, and 10 s were necessary to produce a volume of 20–25 mL for espresso cup, and their caffeine content were of 4.2, 4, and 3.2 mg/mL, respectively [90].

Finally, when people consume espresso coffee at coffee shop, the *barista* can use two types of filter holder at 1 cup or 2 cups. Severini et al. [96], who studied the potential effect of some variables under the control of *barista,* reported that the espresso coffees from 2-cups filter holder presented a higher amount of caffeine. This was explained by the higher extraction of water-soluble compounds as a consequence of the greater amount of coffee ground (~14 g) in the 2-cups filter holder respect to 1 cup (~7 g) [124]. As proved by the authors, during the first 8 s of percolation, the caffeine content in each coffee cup, prepared with fine-coarse (or medium) powder, resulted of 4.51 and 3.46 mg/mL for the beverage prepared with 2-cups and 1-cup filter holder, respectively [96].

On the basis of above discussion, we must state that although the initial choice of coffee blend and the roasting level are important factors affecting the chemical composition of coffee beverage, they are not able to definitively control the amount of caffeine in espresso coffee cup. Many other factors may also counterbalance their effects; likewise, some of these may exhibit an opposite effect. For instance, by increasing the dose of ground coffee, an increase of caffeine content would be expected, but a slight pressure on the surface of coffee cake and a negligible increase of brewing time could increase the amount of water falling in the cup reducing the concentration of caffeine. Therefore, several EC preparation factors should be taken into account contemporaneously, such as the grinding level of coffee powder, the dose, the tamping, the extraction rate, and the volume of extract. Each of them should be precisely defined to obtain the desired chemical compounds and caffeine concentration in EC.

6.4. American coffee

Drip filter or American coffee brew is prepared using an automatic machine (**Figure 6**) [125] equipped of a tank in which the water is heated (92–96°C), a container in which, using a single-use paper filter, is placed the roasted coffee powder. At the bottom of the device, a glass flask collects the coffee beverage. Being the most diffused preparation coffee method in the world, a wide literature on American coffee brew is available. As for other preparation methods, several factors affect the caffeine content in filtered coffee, such as roasting degree, grinding level of ground coffee, dose of coffee, powder/water ratio, brewing time, and final volume of beverage. Tfouni et al. [126] evaluated that the caffeine concentration in filtered coffee brew, obtained by Brazilian coffee beans, roasted at two different levels (medium and dark), varies from 0.92 to 0.99 mg/mL and 1.23–1.65 mg/mL for Arabica and Robusta, respectively. However, another study on the two varieties reported that the amounts of caffeine in American coffee ranged from 0.35 to 1.07 mg/mL and 0.65 to 1.58 mg/mL for Guatemala (Arabica) and Vietnam (Robusta) coffee, respectively. Considering the same extraction time (375 s), the average content of caffeine in filtered coffee brew is about 0.57 and 1.15 mg/mL [101] for Arabica and Robusta, respectively. Bell et al. [93] reported that the finely ground coffee powder yielded a significantly higher caffeine content due to the larger surface area, they highlighted that by using a powder/water ratio of 0.023 g/mL, the caffeine concentration of 0.2, 0.35, and 0.40 mg/mL for coarse, medium, and fine ground coffee were observed, respectively. Also, a longer brewing time (from 3 to 10 min) implies a longer contact time between the water and coffee powder, leading a more complete caffeine extraction.

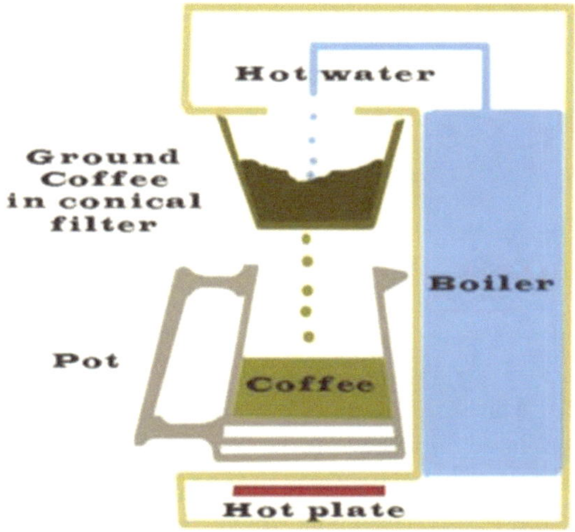

Figure 6. Schematic representation of an American coffee device [125].

6.5. Other coffee preparation methods

As previously reported, apart the preparation of espresso coffee, several other brewing methods may be used to prepare coffee beverages. Among these, some are widely used, popular, and very appreciated in the world, such as filtered coffee, while others are only linked to some cultural tradition and used exclusively in restricted geographical areas. In this section, we summarized the most important brewing method and their specific effects on caffeine content.

The most popular household coffee-brewing method in Italy is the Moka that uses a stove-top coffee maker invented in 1993 by the aluminum technologist, Alfonso Bialetti. Due to its low cost and easy-to-handle characteristics, Moka is used in other countries where it is called stove-top espresso or often misnamed mocha or mocca. In **Figure 7** [127, 128], the moka apparatus is shown, that consists of a metallic tank base, used as a water boiler, a metallic filter to contain the coffee powder, and the cylindrical tank on the upper part in which the coffee brew is collected. The extraction steps are also reported [127]. Boiling water is forced through the filter, containing the coffee ground, up to the tank in which is collected the coffee beverage. Nicoli et al. [1] highlighted that, using a roasted coffee blend and coffee powder/water ratio of 8 g/80 mL, the caffeine content in beverage was 2.56 mg/mL. López-Galilea et al. [100], by using a lower ratio of 40 g/500 mL, reported a caffeine content of 0.28 mg/mL. By using 100% Arabica roasted coffee, with a coffee powder/water ratio of 10 g/50 mL, a high value of caffeine content was found of 5.40 mg/mL [108], while a lower value of 1.68 mg/mL was found employing a ratio of 11.3 g/80 mL [103]. Another study showed that in 100% Arabica coffee brew from moka, with a p/w ratio of 7 g/110 mL, the caffeine content resulted to 0.75 mg/mL [102]. Briefly, exclusively considering the brewing method, the core of moka system is the coffee powder/water ratio used during brew preparation. Of course, when this is reduced, a less

Figure 7. Schematic representation of the use of moka for coffee preparation (adapted from Refs. [127, 128]).

caffeine concentration in the brew is obtained. In these conditions, the total amount of caffeine intake should be linearly related with the volume of brew. Finally, it can be taken into account that not all water is used for coffee brew since that a small part of it remains in the metallic tank base, and another fraction remains in the wet spent coffee.

Another typical Italian method of coffee preparation consists in the use of the Neapolitan pot, also called *cuccumella* that in, Southern Italy, has been very popular. This method is based on the percolation of hot water under gravity through a bed of medium-coarse ground coffee. The *cuccumella* consists of a special coffee pot in aluminum, in which there is a tank filled with water at the bottom, a filter containing the unpressed ground coffee in the middle, and a tank which sealed the upper side of coffee pot on the top. The process consists of heating water in the boiler tank of the coffee pot. When the water reaches the boiling temperature quickly, the Neapolitan machine is overturned, enabling the hot water to percolate across coffee powder and to collect the brew in the upper tank, now down (**Figure 8**) [128, 129]. As reported from Santini et al. [108], 10 g of ground coffee and 50 mL of water are typically used to prepare Neapolitan coffee. According to the limited use of this method, which is restricted in some regional area of Italy, very few studies reported scientific data on the quality of coffee prepared with Neapolitan pot. By using a 100% Arabica roasted coffee, some researches highlighted that a caffeine content of 1.89 mg/mL was measured when using a roasted coffee powder/water ratio of 10 g/50 mL [108], while a value of 1.3 mg/mL was found using a ratio of 15.4 g/145 mL [103]. A coarse coffee powder is necessary to prepare a coffee cup, and after the filtration, the light brown beverage obtained resulted to be very similar to the American coffee.

French coffee, also known as European coffee, is prepared using the French press or plunger pot schematically depicted in **Figure 9** [128]. In this apparatus, the coarse-roasted coffee powder is soaked with hot water for 2 or 5 min, then a separation of ground spent coffee is made pushed down the wire-mesh filter (or plunger) toward the bottom of the tank. Finally, the infused coffee brew may be easily spilled in cup. Also in this case, very few experiments explored the caffeine content of French coffee. López-Galilea et al. [100], who prepared a plunger coffee brew using 40 g of roasted coffee powder and 500 mL of hot water, measured caffeine content of 0.20 mg/mL. Also, Gloess et al. [102] investigated the caffeine concentration of different obtained coffee brews, according to different extraction methods. Among these brewing techniques, the French coffee samples, prepared by 27.5 g of ground coffee in 500 mL of hot water (90°C) for an extraction time of 4 min, exhibited an average caffeine content of 0.49 mg/mL.

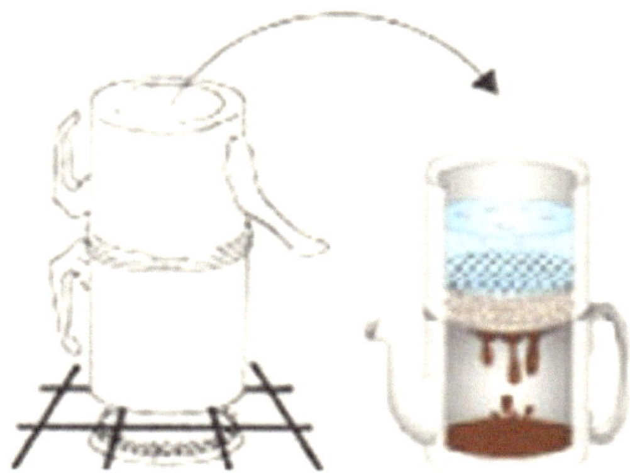

Figure 8. Schematic representation of Neapolitan coffee preparation with Neapolitan pot "Cuccumella" (adapted from Refs. [128, 129]).

Figure 9. Schematic representation of French coffee preparation (adapted from Ref. [128]).

Turkish coffee is the most ancient preparation method of coffee brew. Usually, roasted beans of *C. arabica*, after milling to obtain the finest powder, are boiled in a pot called "*cezve*" (**Figure 10**) [128] previously added with sugar. The coffee is served in a cup where the grounds are allowed to settle. The amount of water necessary for brewing is measured by using the coffee cups but, usually, is the range of 60–90 mL. For each cup, between 5 and 10 g of finest coffee powder are used [106]. A slow heating until the boiling is performed allowing the development of the foam on the beverage surface. Then, the process is interrupted for few seconds before repeating the boiling with the aim to facilitate the precipitation of insoluble compounds. Niseteo et al. [20] reported that the caffeine content of Turkish coffees prepared by using coffee powder/water ratio of 7 g/50 mL, resulted between 2 and 2.8 mg/mL. Similar results were found by Santini et al. [108], who found caffeine content of 1.9 mg/mL, by using 100% Arabica roasted coffee, with a coffee powder/water ratio of 10 g/100 mL.

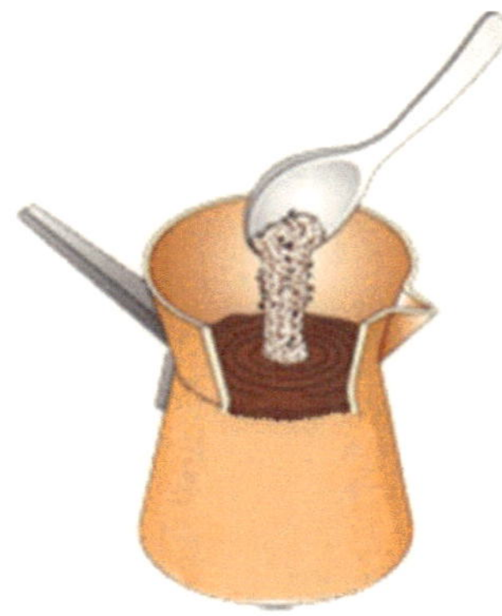

Figure 10. Cezve (adapted from Ref. [128]).

6.6. The use of soluble coffee

Instant, soluble or dried coffee is referred to the soluble portion of roasted-ground coffee, in either powder or granule form, which produces, in a very short time, a coffee beverage adding only hot water to the powder in cup [130]. The production of instant coffee involves the treatment of roasted-ground coffee with hot water and high pressure to extract the water-soluble compounds. Then, the obtained product is subjected to cooling, centrifugation, and concentration by heat and freeze drying at low temperatures [3]. Depending on the coffee species (Arabica or Robusta), roasted degree, and the extraction methods (using hot water or double-extraction, modulating temperature and pressure), different caffeine content may be observed. Vignoli et al. [109] reported that the caffeine content of dark soluble coffee resulted, for both extraction methods, as an average of 3.49 g/100 g and 4.82 g/100 g for Arabica and Robusta soluble coffee, respectively. Niseteo et al. [20] reported that, using a coffee powder/water ratio of 7 g/50 mL, the average caffeine content in two blends of instant coffee was of 4.5 mg/mL. Moreover, by studying eight different brands of instant coffee, Ludwig et al. [110] highlighted that the amount of caffeine content in coffee brews, prepared by 2 g of instant coffee dissolved in 125 mL of boiling water, ranged from 0.38 to 0.70 mg/mL, with an average value of 0.46 mg/mL.

6.7. Single service size systems: pods and capsule

In the last decade, a new coffee preparation method was developed to fulfill the increasing needs of consumers, such as convenience, high quality, quickness, and ease of use. The roasted coffee powder is dosed, tamped, and hermetically packaged following two methods: (1) pods, obtained by sealing the ground coffee between two layers of filter paper; (2) capsules of different size and shape but essentially in plastic or aluminum. The key factor of their success is to make possible for anyone to prepare a like-espresso coffee anytime and everywhere. However, the use of pods or capsules shows great differences, and each coffee brand has developed brewer machines with specific features, such as pressure, percolation time, water temperature, flow rate of water, etc. to obtain a quality of coffee as best as possible.

http://dx.doi.org/10.5772/intechopen.69002

Several studies reported on the use of capsule or pods to make an espresso coffee cup. Albanese et al. [111] studied five blends of roasted coffee (100% Arabica (A), 100% Robusta (R), 80% A–20% R, 40% A–60% R, and 20% A–80% R) packaged in pods and extracted by three water temperature (90, 100, and 110°C) and their effects on chemical properties of espresso coffee brews (coffee powder/water ratio: 7 g/25 mL). As expected, the caffeine content was strictly depended from coffee blend; in fact, increasing the percentage of Robusta coffee, the caffeine content in the extracts resulted higher. In addition, the high temperature of water promoted the extraction of chemical compounds among which the caffeine. As reported, the caffeine contents in ECs were of 2.59 mg/mL (100% A) and 3.55 mg/mL (100% R) when extracted at water temperature of 90°C and 3.31 mg/mL (100% A) and 4.65 mg/mL (100% R), when extracted at water temperature of 110°C.

Bartel et al. [131] studied several single-service systems (pods and capsules) to prepare espresso coffee samples. The caffeine content for an espresso "*lungo*" (100% A) prepared by pods (coffee powder/water ratio: 6.9 g/115 mL) was 0.79 mg/mL, while similar values of 0.80 and 0.77 mg/mL were measured using plastic or aluminum capsules, respectively. However, it must be considered that the use of the above three systems implies the use of different extraction conditions with coffee powder/water ratios of 6.9 g/115 mL, 7.9 g/115 mL, and 5.2 g/85 mL for pods, plastic capsule, and aluminum capsules, respectively. Obviously, differences were also found in caffeine content for EC from coffee blend (35% A–62% R; coffee powder/water ratio: 5.2 g/85 mL) in aluminum capsule with a mean value of 1.08 mg/mL, while, Gloess et al. [102] reported that the caffeine content in a regular EC, from aluminum capsule (100% A; coffee powder/water ratio: 5.5 g/30 mL), resulted of 1.4 mg/mL.

A recent research reported that, using the same roasted coffee powder, comparing two single-dose capsule systems to the classic bar machine, the caffeine content in ECs resulted equivalent, having an average value of 2.22 mg/mL [112].

Another study observed that, keeping constant the particle size distribution, the pressure on the upper surface of coffee cake (i.e. the tamping) in different brands of single-dose capsule, may have an important effect on the extraction of caffeine in ECs due to the changes in microstructure of coffee cake [113].

A complex research on single-serve capsule brewer to prepare the American coffee highlighted that several parameters, such as the origin of raw material, the roasting degree, the particle size distribution of coffee powder, the dose in capsule and the cup volume, significantly affected the chemical and sensorial attributes of coffee brews [104]. Considering constant some variables, as the dose of coffee (8.9 g) and the grinding level (volume mean diameter = 734 μm), the caffeine content in American coffee brews increased when the roasting level was high (dark > medium) and resulted lower when the volume of beverage increased from 113 to 226 mL.

Of course, independently on the extraction system used, classic coffee machine (i.e. Espresso or American) or single-dose systems, the same variables affect the chemical and sensorial properties of coffee brew, such as the grinding level, the dose of powder, the tamping (Espresso), the extraction time, and the volume of beverage.

7. How much caffeine in a single cup? Differences through brewing method and conditions

How much caffeine is actually assumed for coffee cup? Even though it is recognized that 50 mg of caffeine for cup and 4 cups/day (total amount of 200 mg/day) is acceptable for people, a real assessment of caffeine intake for consumers is very difficult. It is a non-trivial question in consideration of that it depends on the brew volume (i.e. how big is the cup), the grinding grade, the dose, the tamping, the brewing method used, how much coffee ground is used to prepare the brew, the coffee varieties, and blending. Of course, this could become a problem when considering that each people cannot known the total content of caffeine inside a cup consumed at home, at coffee shops, by self-service coffee machine, etc. Crozier et al. [12] did a snapshot of the variability of caffeine content of espresso coffees sold in several coffee shops. They reported that caffeine may vary of 6-fold from 51 mg/cup in Starbucks to 322 mg/cup in Patisserie Francoise. This impressive variability is certainly the result of different extraction conditions, mainly the dose but also grinding level, roasting conditions, volume of coffee cup, etc.

The web site caffeine informer [132] enables to examine the content in caffeine of hundreds of coffee brews sold by different brands. By sorting in ascending order, the first one is the Nescafe Ice Java having a caffeine content of 100 mg in 25 mL (4 mg/mL), while the last one is, as expected, the decaffeinated instant coffee with 2.5 mg in 236 mL (0.0106 mg/mL). In **Table 4** [132], the amount of caffeine for the most popular coffee brews and the related volume are resumed.

McCusker et al. [18] analyzing the caffeine content of "speciality" reported a great variability among coffee types as well as among coffees sold in different days but in the same coffee bar. As example, they reported caffeine doses of 75.8 mg and 140.4 mg for 1-shot (42 mL) and for 2 short shots (40 mL) of espresso coffee respectively, while, when a 1-shot of coffee (42 mL) was sold by Starbucks, a significantly lower content in caffeine of 58.1 mg was measured. Similarly, Crozier et al. [12], by considering espresso coffees sold by Starbucks, showed caffeine content of 51 mg for a serving size of 27 mL. This first data clearly indicate a great variability in caffeine content in cup among the brewing methods, total brew volume in cup as well as inside the same coffee shop.

McCusker et al. [18], analyzing some brands specialty coffees in a 16-oz cup (473 mL), reported caffeine content between 143.4 and 259.3 mg respectively for Dunkin' Donuts and Starbucks. A very interesting finding was the high variability observed by analyzing Starbucks' coffees, although it was expect a high standardization of preparation conditions. Particularly, analyzing six consecutive days, the authors reported caffeine contents between 259.2 and 564.4 mg for a 473 mL cup. Furthermore, by consulting the website of Starbucks [133] caffeine content of 155, 235, 319, and 410 mg are, reported for a short 8-oz cup (236 mL), a "tall" coffee of 12-oz (354.72 mL), a "grande" coffee of 16-oz (472.96 mL), and a "venti" coffee of 20-oz (591.2 mL), respectively. By plotting the caffeine contents vs volume, we evaluate a direct linear relationship between volume and caffeine content but this was not confirmed by scientific literature. This was because in the coffee shop like Starbuck's, coffee brew is continuously prepared and stored in big urn until where, of course, caffeine content is an average of several

Coffee brew	mL	Caffeine (mg)	mg/mL
Nescafe Ice Java	25.14	100	3.98
Black Insomnia Coffee	354.88	702	1.98
Coffee (Espresso)	44.36	77	1.74
Nespresso Coffee Capsules	39.92	60	1.50
Robusta Coffee	236.59	265	1.12
Turkish Coffee	59.15	50	0.85
Illy Issimo Cafe	201.10	155	0.77
Starbucks Grande Coffee	473.18	330	0.70
High Brew Coffee	236.59	163	0.69
Starbucks Doubleshot	192.23	125	0.65
Dunkin' Donuts Brewed Coffee	414.03	210	0.51
Starbucks Grande Caffe Americano	473.18	225	0.48
Americano Coffee	354.88	154	0.43
Caffe Mocha	354.88	152	0.43
Starbucks Grande Caffe Mocha	473.18	175	0.37
McDonalds (McCafe) Mocha	473.18	167	0.35
McDonalds Coffee	473.18	145	0.31
Starbucks Verismo Coffee Pods	236.59	60	0.25
Coffee (Instant)	236.59	57	0.24
Caffe Nero Coffee	354.88	80	0.23
Starbucks Decaf Coffee	473.18	25	0.05
Nescafe' Ricoffy	236.59	6	0.03
Coffee (Decaf, Brewed)	236.59	6	0.03
Coffee (Decaf, Instant)	236.59	3	0.01

Table 4. Caffeine content in cup for coffee brews sold by several brands [132].

extractions. In these conditions, the average caffeine concentration will be exactly the same for each coffee cup, while the only significant variable becomes the total volume of beverage.

Ludwig et al. [101] studied the effect of brewing time and two different methods, such as espresso and filtered coffee brews, on caffeine content. Filtered coffee brews were prepared by using 36 g of ground coffee, 600 mL of water at 90°C, and a brewing time of 6 min. Espresso coffee samples were obtained employing a conventional coffee machine from 7 g of ground coffee for a brew volume of 45 mL. The authors separated coffee samples in 5 and 3 fraction for espresso and filtered coffee, respectively, and they analyzed the changes in caffeine concentration (mg/10 mL) and volumes (mL) of each brew fraction. For instance, caffeine content reduced from 85.44 mg (106.8 mg/100 mL) in the first fraction (75 s, 80 mL) to 23.14 mg (89 mg/100 mL) in the fifth fraction (75 s, 26 mL) in the case of filtered coffee. On the other hand, values from 47.50 mg (in the first fraction of 8 s and 16 mL) to 5.03 mg (for the third fraction of 8 s and 17 mL) were reported for espresso coffee. By using these data, it was possible to calculate that, for people consuming a 473 mL of coffee cup (i.e. a part of the total volume of 600 mL), a caffeine intake of 304.02 and 545.36 mg should be considered for filtered brew prepared by Guatemala and Vietnam coffee, respectively. Similarly, values of 63.63 and 131.98 mg were found assuming 45 mL of espresso, when using Guatemala and Vietnam coffee, respectively. These data are in accordance to that reported by McCusker et al. [18] for 1-shot of espresso coffee, while they were significantly higher for filtered coffee.

Parenti et al. [112] reported the differences in caffeine content of espresso coffee comparing different brewing techniques. The authors compared the espresso coffees from conventional bar machine, the hyper espresso method (HIP) and the I-Espresso (IE) capsule systems, reporting a total volume of 25–30 mL (with a flow rate of 1 mL/s) for conventional bar machine, while, for HIP and IE, the volume of EC brews was weighed until 25 g. However, if let us consider that the authors prepared a regular coffee of 25 mL for each type of brewing method, we estimated values of 55.5, 57.75, and 53.5 mg of caffeine for conventional bar machine, hyper espresso and IE systems, respectively. These data are lower than those previously discussed from Ludwig et al. [101], who used a lower total volume (25 vs. 45 mL) and a less amount of ground coffee (6.7 vs. 7 g).

Caporaso et al. [103] analyzed the caffeine content of Neapolitan, Moka, Espresso, and American (filtered) coffee. For espresso coffee (25 mL), a caffeine content of 60.95 mg for cup was obtained, being in good agreement with the findings of Parenti et al. [112]. On the other hand, no accordance there was for the American coffee samples for which the authors measured a dose of caffeine of 173.25 mg that is significantly lower than the data reported from both McCusker et al. [18] and Ludwig et al. [101]. However, this was mainly due to the changes in total volume of the brew that in the case of the paper of Caporaso and coauthors. was considered of 125 mL. By considering the caffeine concentration of 1.39 mg/mL as reported from the authors, it is possible to estimate a total content of caffeine for a 16-oz cup (473 mL) of 657.47 mg. This value is greater than those reported from McCusker et al. [18] and Ludwig et al. [101] as above reported for the same volume of coffee cup.

Relationship between caffeine content and four brewing procedures (filter, plunger, mocha, and espresso coffee method) were also studied by López-Galilea et al. [100]. Considering a commercial blend of Arabica and Robusta, caffeine concentrations of 0.22, 0.20, 0.28, and 0.63 mg/mL were measured for filter, plunger, mocha, and espresso coffee, respectively. From these data, total caffeine of 25, 140, 100, and 88 mg may be estimated for espresso (40 mL), mocha (500 mL), plunger (500 mL), and filtered coffee (400 mL), respectively.

All above data indicates two main aspects: (1) a very high variability of caffeine content in cup, also when we consider the same brewing conditions and (2) the difficulty to critically compare the literature data since they are obtained in different operative conditions. Particularly, for the latter consideration, the kinetic of caffeine extraction should be always taken into account by the researchers, who wish give information on caffeine intake as well as for each other chemical compounds. This is because the caffeine kinetic extraction is not linearly related with time. For instance, if we consider the data published by López-Galilea et al. [100], an average caffeine concentration of 0.22 mg/mL was measured for filtered coffee, leading to a total content of 88 mg in a 400 mL of total volume. However, the comparison of these results and those reported by McCusker et al. [18] and Ludwig et al. [101] who analyzed a 16-oz cup (473 mL) is not possible. This is because, in both papers any information on the kinetic of caffeine extraction were not reported. Severini et al. [90] studied how the variance of some extraction variable may affect the quality of espresso coffees served every day. The authors modeled the kinetic of caffeine extraction by changing the grinding (coarse, fine-coarse, and fine ground coffee), the dose (6, 7, and 8 g), and the tamping on the upper surface of coffee cake (0.75, 1.5, and 2.25 kg). **Figure 11** [90] reports the kinetic extraction of caffeine and its cumulative dose as a function of extraction time for sample prepared by coarse (grinding level, 7), fine-coarse (grinding level, 6.5), and fine ground coffee (grinding level, 6). The authors estimated total caffeine contents of a 25 mL cup as 77.4, 105.83, and 98.97 mg for brew prepared by coarse, fine-coarse, and fine ground coffee, respectively.

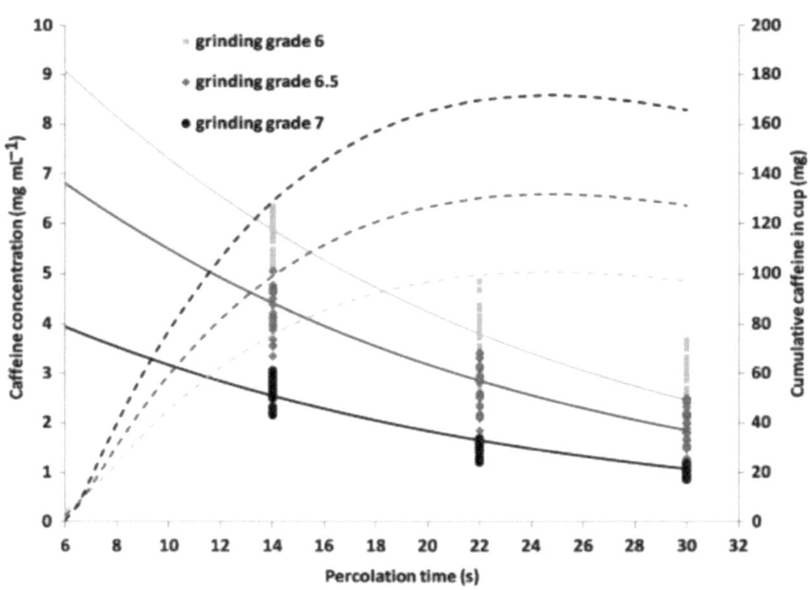

Figure 11. Changes in caffeine concentration of espresso coffee as a function of grinding level and extraction time [90].

8. Conclusion

From the huge number of researches and results, which we can find in literature, it becomes quite impossible to answer a simple question: how much caffeine we take with a cup of coffee?

Different cultivar, origins, agronomic conditions, post-harvest treatments, transport and conservation, as well as the blending before the roasting could affect the caffeine content in green coffee seeds.

The roasting process seems to be the only step almost irrelevant, because the caffeine remains more or less unaltered by the roasting temperature.

Each different operative condition, such as grinding level, dose of ground coffee, tamping, water temperature, water pressure, water/coffee ratio, extraction rate, volume of beverage, etc. could produce differences in the extraction kinetic of caffeine which should be considered when comparing the caffeine content in cups. Unfortunately, despite the wide bibliography concerning the caffeine content in coffee brew, few papers reported the differences in extraction kinetic of caffeine by changing type of coffee brewing.

Among all process parameters, doubtless the grinding level plays an important role for the caffeine content "in cup", due to its effect on extraction kinetic. However, the considerable variability in the composition of the coffee beverage, as well as the significant differences in volume of a single coffee cup, makes it very hard to accurately define the average of daily intake of caffeine and of other bioactive constituents of coffee.

From the point of view of caffeine effects on human health, its content in coffee cup and its intake should be far to be a trouble. In the same way, it seems unjustified the choice of a pure variety of green coffee, based on the less content in caffeine.

Despite 20 years of reassuring researches, many people still avoid caffeinated coffee because they worry for the biological effects of caffeine [10].

A difficulty in interpreting epidemiological data is that some surveys were not specifically designed to quantify coffee consumption; thus, the debate about the coffee consumption, its beneficial or detrimental effect for human health, still persists. Pending that these encouraging observations could be confirmed and be widely spread, further experiments are needed particularly on the bioavailability of coffee components in order to elucidate their responsibility as well as the mechanisms involved in the observed positive effects. It may be concluded, therefore, that labeling the coffee as a harmful beverage and caffeine as a dangerous compound for human health lacks of support in the light of present knowledge.

Acknowledgements

The research was carried out from the Authors thanks to the support of ESSSE Caffè S.p.A. (Anzola dell'Emilia, BO, Italy).

Author details

Carla Severini*, Antonio Derossi, Ilde Ricci, Anna Giuseppina Fiore and Rossella Caporizzi

*Address all correspondence to: carla.severini@unifg.it

Department of Science of Agriculture, Food and Environment, University of Foggia, Italy

References

[1] Nicoli MC, Dalla Rosa M, Lerici CR, Severini C. Caratteristiche chimiche dell'estratto di caffè. Nota 1: Cinetiche di estrazione della caffeina e delle sostanze solide. Industrie Alimentari. 1987;**26**:467-471

[2] Clarke RJ. Coffee: Roast and Ground. Encyclopedia of Food Sciences and Nutrition. 2nd ed; 2003. pp. 1487-1493. Available from: http://lenscoffee.com [Accessed: 21-02-2017]

[3] Farah A. Coffee constituent. In: Chu YF, editor. Coffee: Emerging Health Effects and Disease Prevention. Oxford: Blackwell Publishing Ltd; 2012; 352. ISBN: 978-0-470-95878-0

[4] Bicho NC, Lidon FC, Ramalho JC, Leitao AE. Quality assessment of Arabica and Robusta green and roasted coffees—A review. Emirates Journal of Food and Agriculture. 2013;**25**:945-950. DOI: 10.9755/ejfa.v25i12.17290

[5] Babova O, Occhipinti A, Maffei M. Chemical partitioning and antioxidant capacity of green coffee (*Coffea arabica* and *Coffea canephora*) of different geographical origin. Phytochemistry. 2016;**123**:33-39. DOI: 10.1016/j.phytochem.2016.01.016

[6] Regulation (EC) No 1924/2006 of the European Parliament and of the Council of 20 December 2006 on Nutrition and Health Claims Made on Foods

[7] Coughlin JR, Nehlig A. Coffee and cancer: A benefit-risk evaluation of the experimental and epidemiological evidence. In: The 24th International Conference on Coffee Sciences (ASIC 2012)-12-16 November, San José, Costa Rica. ASIC Pub. - Wädenswil (Switzerland) pp. 25-32

[8] Linn S, Schoenbaum SC, Monson RR, Rosner B, Stubblefield PG, Ryan KJ. No association between coffee consumption and adverse outcomes of pregnancy. New England Journal of Medicine. 1982;**306**:141-145. DOI: 10.1056/NEJM198201213060304

[9] Devasagayam TP, Kamat JP, Mohan H, Kesavan PC. Caffeine as an antioxidant: Inhibition of lipid peroxidation induced by reactive oxygen species. Biochimica et Biophysica Acta. 1996:**1282**(1):63-70. DOI: 10.1016/0005-2736(96)00040-5

[10] Bae KH, Park JH, Imm SS, Song DK. Coffee and health. Integrative Medicine Research. 2014;**3**:189-191. DOI: 10.1016/j.imr.2014.08.002

[11] ICO. Global Coffee Consumption. 2016. Available from: http://www.ico.org/monthly_coffee_trade_stats.asp [Accessed: 11-11-2016]

[12] Crozier TWM, Stalmach A, Lean MEJ, Crozier A. Espresso coffees, caffeine and chlorogenic acid intake: Potential health implications. Food & Function. 2012;**3**(1):30-33. DOI:10.1039/c1fo10240k

[13] Bonita JS, Mandarano M, Shuta D, Vinson J. Coffee and cardiovascular disease: In vitro, cellular, animal, and human studies. Pharmacological Research. 2007;**55**:187-198. DOI: 10.1016/j.phrs.2007.01.006

[14] NCA. National Coffee Drinking Trends 2012. National Coffee Association; 2012 U.S.A

[15] Spiller MA. The chemical components of coffee. Progress in Clinical and Biological Research. 1984;**158**:91-147. DOI: 10.1201/9781420050134.ch6

[16] Gonzalez de Mejia E, Ramirez-Mares MV. Impact of caffeine and coffee on our health. Trends in Endocrinology and Metabolism. 2014;**25**(10):498-592. DOI: 10.1016/j.tem.2014.07.003

[17] Xia Z, Ni Y, Kokot S. Simultaneous determination of caffeine, theophylline and theobromine in food samples by a kinetic spectrophotometric method. Food Chemistry. 2013;**141**:4087-4093. DOI: 10.1016/j.foodchem.2013.06.121

[18] McCusker RR, Goldberger BA, Cone EJ. Caffeine content of specialty coffees. Journal of Analytical Toxicology. 2003;**27**:520-522

[19] Heckmann MA, Weil J, Gonzalez de Mejia E. Caffeine (1,3,7-trimethylxanthine) in foods: A comprehensive review on consumption, functionality, safety and regulatory matters. Journal of Food Science. 2010;**75**:R77-R87. DOI: 10.1111/j.1750-3841.2010.01561.x 10.1111

[20] Niseteo T, Komes D, Belscak-Cvitanovic A, Horzic D, Budec M. Bioactive composition and antioxidant potential of different commonly consumed coffee brews affected by their preparation technique and milk addiction. Food Chemistry. 2012;**134**:1870-1877. DOI: 10.1016/j.foodchem.2012.03.095

[21] Frary CD, Johnson RK, Wang MQ. Food sources and intakes of caffeine in the diets of persons in the United States. Journal of the American Dietetic Association. 2005;**105**:110-113. DOI: 10.1016/j.jada.2004.10.027 10.1016

[22] Mitchell DC, Knight CA, Hockenberry J, Teplansky R, Hartman TJ. Beverage caffeine intakes in the U.S. Food and Chemical Toxicology. 2014;**63**:136-142. DOI: 10.1016/j.fct.2013.10.042

[23] U.S. Food and Drug Administration. Letter from Jeanne Ireland, Assistant Commissioner for legislation, FDA, to Senator Richard J. Durbin (August 10, 2012). (http://www.durbin.senate.gov/public/index.cfm/files/serve?File_id=17eadaa1-85e7-4ceb-a827-be244fbddfa5)

[24] Health Canada. Caffeine, it's Your Health. 2010 (http://www.hc-sc.gc.ca/fn-an/securit/addit/caf/index-eng.php; [Accessed: 10-01-2017]

[25] Brent, RL, Christian MS, Diener RM. Evaluation of the reproductive and developmental risks of caffeine. Birth Defects Research Part B: Developmental and Reproductive Toxicology. 2011;**92**:152-187. DOI: 10.1002/bdrb.20288

[26] Peck, JD, Leviton A, Cowan LD. A review of the epidemiologic evidence concerning the reproductive health effects of caffeine consumption: A 2000-2009 update. Food and Chemical Toxicology. 2010;**48**:2549-2576. DOI: 10.1016/j.fct.2010.06.019

[27] Cano-Marquina A, Tarin JJ, Cano A. The impact of coffee on health. Maturitas. 2013;**75**:7-21. DOI: 10.1016/j.maturitas.2013.02.002

[28] Borota D, Murray E, Keceli G, Chang A, Watabe JM, Ly M, Toscano JP, Yassa MA. Post-study caffeine administration enhances memory consolidation in humans. Nature Neuroscience. 2014;**17**(2):201-203. DOI: 10.1038/nn.3623

[29] Rudelle S, Ferruzzi MG, Cristiani I, Moulin J, Macè K, Acheson KJ, Tappy L. Effect of a thermogenic beverage on 24-hour energy metabolism in humans. Obesity. 2007;**15**:349-355. DOI: 10.1038/oby.2007.529

[30] Frizzarin R, Maya F, Estela JM, Cerdà V. Fully-automated in-syringe dispersive liquid-liquid microextraction for the determination of caffeine in coffee beverages. Food Chemistry. 2016;**212**:759-767. DOI:10.1016/j.foodchem.2016.06.032

[31] Ludwig IA, Clifford MN, Lean MEJ, Ashihara H, Crozier A. Coffee: Biochemistry and potential impact on health. Food & Function. 2014;**5**:1695-1717. DOI: 10.1039/c4fo00042k

[32] Rogers P, Hohoff C, Heatherley SV, Mullings EL, Maxfield PJ, Evershed RP, Deckert J, Nutt DJ. Association of the anxiogenic and alerting effects of caffeine with ADORA2A and ADORA1 polymorphisms and habitual level of caffeine consumption. Neuropsychopharmacology. 2010;**35**:1973-1982. DOI: 10.1038/npp.2010.71

[33] Lucas M, Mirzaei F, Pan A, Okereke OI, Willets WC, O'Reilly RJ, Koene K, Ascherio A. Coffee, caffeine and depression among women. Archives of Internal Medicine. 2011;**171**(17):1571-1578. DOI:10.1001/archinternmed.2011.393

[34] Williamson G. Coffee and health. In: Crozier H, Ashihara H, Tomas-Barberan F, editors. Teas, Cocoa and Coffee: Plant Secondary Metabolites and Health; 2012. Oxford: Wiley-Blackwell; pp. 169-181

[35] Nawrot P, Jordans S, Eastwood J, Rotstein J, Hugenholtz A, Feeley M. Effects of caffeine on human health. Food Additives and Contaminants. 2003;**20**:1-30. DOI: 10.1080/0265203021000007840

[36] Huxley R, Lee M, Barzi F, Timmermeister L, Czernichow S, Perkovic V, Grobbee DE, Batty D, Woodward M. Coffee, decaffeinated coffee, and tea consumption in relation to incident type 2 diabetes mellitus A systematic review with meta-analysis. Archives of Internal Medicine. 2009;**169**:2053-2063. DOI: 10.1001/archinternmed.2009.439 10.1001

[37] Hidgon JV, Frei B. Coffee and health: A review of recent human research. Critical Reviews in Food Science and Nutrition. 2006;**46**:101-123. DOI: 10.1080/10408390500400009

[38] Johnston KL, Clifford MN, Morgan LM. Coffee acutely modifies gastrointestinal hormone secretion and glucose tolerance in humans: Glycemic effects of chlorogenic acid and caffeine. American Journal of Clinical Nutrition. 2003;**78**:728-733

[39] Greenberg JA, Owen DR, Geliebter A. Decaffeinated coffee and glucose metabolism in young men. Diabetes Care. 2010;**33**:278-280. DOI: 10.2337/dc09-1539

[40] Homan DJ, Mobarhan S. Coffee: Good, bad, or just fun? A critical review of coffee's effects on liver enzymes. Nutrition Reviews. 2006;**64**:43-46. DOI: 10.1111/j.1753-4887.2006.tb00172.x

[41] Salomone F, Li Volti G, Vitaglione P, Morisco F, Fogliano V, Zappalà A, Palmigiano A, Garozzo D, Caporaso N, D'Argenio G, Galvano F. Coffee enhances the expression of chaperones and antioxidant proteins in rats with nonalcoholic fatty liver disease. Translational Research. 2014;**163**(6):593-602. DOI: 10.1016/j.trsl.2013.12.001

[42] Masterton GS, Hayes PC. Coffee and the liver: A potential treatment for liver disease? European Journal of Gastroenterology & Hepatology. 2010;**22**:1277-1283. DOI: 10.1097/MEG.0b013e32833cca96

[43] Trevitt J, Kawa K, Jalali A, Larsen C. Differential effects of adenosine antagonists in two models of parkinsonian tremor. Pharmacology, Biochemistry, and Behavior. 2009;**94**:24-29. DOI: 10.1016/j.pbb.2009.07.001

[44] Hernan MA, Takkouche B, Caamano-Isorna F, Gestal-Otero JJ. A meta-analysis of coffee drinking, cigarette smoking, and the risk of Parkinson's disease. Annals of Neurology. 2002;**52**:276-284. DOI: 10.1002/ana.10277

[45] Ascherio A, Chen H, Schwarzschild MA, Zhang SM, Colditz GA, Speizer FE. Caffeine, postmenopausal estrogen and risk of Parkinson's disease. Neurology. 2003;**60**:790-795. DOI: 10.1212/01.WNL.0000046523.05125.87

[46] de Mendonca A, Cunha RA. Therapeutic opportunities for caffeine in Alzheimer's disease and other neurodegenerative disorders. Journal of Alzheimer's Disease. 2010;**20**(1):S1-S2. DOI: 10.3233/JAD-2010-01420

[47] Arendash GW, Schleif W, Rezai-Zadeh K, Jackson EK, Zacharia LC, Cracchiolo JR, Shippy D, Tan J. Caffeine protects Alzheimer's mice against cognitive impairment and reduces brain beta-amyloid production. Neuroscience. 2006;**142**:941-952. DOI: 10.1016/j.neuroscience.2006.07.021

[48] Cao C, Wang L, Lin X, Mamcarz M, Zhang C, Bai G, Nong J, Sussman S, Arendash G. Caffeine synergizes with another coffee component to increase plasma GCSF: Linkage to cognitive benefits in Alzheimer's mice. Journal of Alzheimer's Disease. 2011;**25**:323-335. DOI: 10.3233/JAD-2011-110110

[49] Gelber RP, Petrovitch H, Masaki KH, Ross GW, White LR. Coffee intake in midlife and risk of dementia and its neuropathologic correlates. Journal of Alzheimer's Disease. 2011;**23**:607-615. DOI: 10.3233/JAD-2010-101428

[50] Bravi F, Boscetti C, Tavani A, Bagnardi V, Gallus S, Negri E, Franceschi S, La Vecchia C. Coffee drinking and hepatocellular carcinoma risk: A meta-analysis. Hepatology. 2007;**46**:430-435. DOI: 10.1002/hep.21708

[51] Larsson SC, Wolk A. Coffee consumption and the risk of liver cancer: A meta-analysis. Gastroenterology. 2007;**132**:1740-1745. DOI: 10.1053/j.gastro.2007.03.044

[52] Arab L. Epidemiologic evidence on coffee and cancer. Nutrition and Cancer. 2010;**62**:271-283. DOI: 10.1080/01635580903407122

[53] Jiang W, Wu Y, Jiang X. Coffee and caffeine intake and breast cancer risk: An update dose-response meta-analysis of 37 published studies. Gynecologic Oncology. 2013;**129**:620-629. DOI: 10.1016/j.ygyno.2013.03.014

[54] Jee SH, He J, Appel LJ, Whelton PK, Suh I, Klag MJ. Coffee consumption and serum lipids: A meta-analysis of randomised controlled clinical trials. American Journal of Epidemiology. 2001;**153**:353-362. DOI: 10.1093/aje/153.4.353

[55] Riksen NP, Smits P, Rongen GA. The cardiovascular effects of methylxanthines. Handbook of Experimental Pharmacology. 2011;**200**:413-437. DOI: 10.1007/978-3-642-13443-2_16

[56] Dixit SBA, Stein PK, Dewland TA, Dukes JW, Vittinghoff E, Heckbert SR, Marcus GM. Consumption of caffeinated products and cardiac ectopy. Journal of the American Heart Association. 2016;5:1-10. DOI: 10.1161/JAHA.115.002503

[57] Klatsky AL, Hasan AL, Armstrong MA, Udaltsova N, Morton C. Coffee, caffeine, and risk of hospitalization for arrhythmias. The Permanente Journal. 2011;**15**(3):19-25

[58] Dorea JG, da Costa THM. Is coffee a functional food? British Journal of Nutrition. 2005;**93**:773-782. DOI: 10.1079/BJN20051370

[59] Mazzafera P, Crozier A, Magalhaes C. Caffeine metabolism in *Coffea arabica* and other species of coffee. Phytochemistry. 1991;**30**:3913-3916. DOI: 10.1016/0031-9422(91)83433-L

[60] FAS/USDA. Coffee: World Markets and Trade. Office of global analysis, Foreign Agricultural Service/United States Department of Agriculture; 2013

[61] Gielissen R, Graafland J. Concepts of price fairness: Empirical research into the Dutch coffee market. Business Ethics: A European Review. 2009;**18**:165-178. DOI: 10.1111/j.1467-8608.2009.01555.x

[62] Koshiro Y, Zheng X-Q, Wang M-L, Nagai C, Ashihara H. Changes in content and biosynthetic activity of caffeine and trigonelline during growth and ripening of *Coffea*

arabica and *Coffea canephora* fruits. Plant Science. 2006;**171**(2):242-250. DOI: 10.1016/j.plantsci.2006.03.017

[63] Cheng B, Furtado A, Smyth HE, Henry R. Influence of genotype and environment on coffee quality. Trends in Food Science & Technology. 2016;**57**:20-30. DOI:10.1016/j.tifs.2006.09.003

[64] Vaast P, Bertrand B, Perriot JJ, Guyot B, Genard M. Fruit thinning and shade improve bean characteristics and beverage quality of coffee (*Coffea arabica* L.) under optimal conditions. Journal of the Science of Food and Agriculture. 2006;**86**(2):197-204. DOI: 10.1002/jsfa.2338

[65] Odeny D, Chemining'wa G, Shibairo S. Beverage quality and biochemical components of shaded coffee. In: Paper Presented at the 25th International Conference of Coffee Science; 2014, ASIC Pub. - Wädenswil (Switzerland) Colombia

[66] Kumar A, Simmi P, Naik GK, Giridhar P. RP-HPLC and transcript profile indicate increased leaf caffeine in Coffea canephora plants by light. Journal of Biology and Earth Sciences. 2015;**5**(1):1-9. ISSN: 2084-3577

[67] Avelino J, Barboza B, Araya JC, Fonseca C, Davrieux F, Guyot B, Cilas C. Effects of slope exposure, altitude and yield on coffee quality in two altitude terroirs of Costa Rica, Orosi and Santa Maria de Dota. Journal of the Science of Food and Agriculture. 2005;**85**(11):1869-1876. DOI: 10.1002/jsfa.2188

[68] Fox GP, Wu A, Yiran L, Force L. Variation in caffeine concentration in single coffee beans. Journal of Agricultural and Food Chemistry. 2013;**61**:10772-10778. DOI: 10.1021/jf4011388

[69] Coffee Roast Levels. Available from: http://lenscoffee.com [Accessed: 21-02-2017]

[70] Münchow M. Roast Profile Analysis. Available from: (https://coffee-mind.com/profile/; [Accessed: 08-02-2017]

[71] Coste R. Coffee, the Plant and the Product. 10th ed. London: The MacMillan Press; 1992. 328 p. ISBN 10: 0333572963

[72] Strezov V, Evans TJ. Thermal analysis of the reactions and kinetics of green coffee during roasting. International Journal of Food Properties. 2005;**8**:101-111. DOI: 10.1081/JFP-200048060

[73] Dutra ER, Oliveira LS, Franca AS, Ferraz VP, Afonso RJCF. A preliminary study on the feasibility of using the composition of coffee roasting exhaust gas for the determination of the degree of roast. Journal of Food Engineering. 2001;**47**:241-246. DOI: 10.1016/S0260-8774(00)00116-3

[74] Franca AS, Mendonca CFJ , Oliveira SD. Composition of green and roasted coffees of different cup qualities. LWT. 2005;**38**:709-715. DOI: 10.1016/j.lwt.2004.08.014

[75] de Oliveira AL, Cruz PM, Eberlin MN, Cabral FA. Brazilian roasted coffee oil obtained by mechanical expelling: Compositional analysis by GC-MS. Ciênc Tecnol Aliment, Campinas. 2005;**4**:677-682. DOI: 10.1590/S0101-20612005000400009

[76] Tarka SMJ. The toxicology of cocoa and methylxanthines: A review of the literature. CRC Critical Reviews in Toxicology. 1982;**9**:275. DOI: 10.3109/10408448209037495

[77] Toci A, Farah A, Trugo LC. Effect of decaffeination using dichloromethane on the chemical composition of arabica and robusta raw and roasted coffees. Quim Nova. 2006;**29**:965-971. DOI: 10.1590/S0100-40422006000500015

[78] Macrae R. Nitrogenous components. In: Clarke RJ, Macrae R, editors. Coffee: Chemistry. London: Elsevier Applied Science Publishers; 1989; Vol. 1. 306 p. ISBN: 10: 0853343683-ISBN-13: 978-0853343684

[79] Casal S, Oliveira MB, Ferreira MA. HPLC/diode-array applied to the thermal degradation of trigonelline, nicotinic acid and caffeine in coffee. Food Chemistry. 2000;**68**:481-485. DOI: 10.1016/S0308-8146(99)00228-9

[80] Dorfner R, Ferge T, Yeretzian C, Kettrup A, Zimmermann R. Laser mass spectrometry as on-line sensor for industrial process analysis: Process control of coffee roasting. Analytical Chemistry. 2004;**76**:1386-1402. DOI: 10.1021/ac034758n

[81] Viet LT, Truchement B. Swiss Patent CH 665 754 A5; 1998

[82] Nebesny E, Budryn G. Antioxidative activity of green and roasted coffee beans as influenced by convection and microwave roasting methods and content of certain compounds. European Food Research and Technology. 2003;**217**:157-163. DOI 10.1007/s00217-003-0705-4

[83] Nebesny E, Budryn G, Kula J , Majda T. The effect of roasting method on headspace composition of robusta coffee bean aroma. European Food Research and Technology. 2007;**225**:9-19. DOI 10.1007/s00217-006-0375-0

[84] Illy A. Quality. In: Illy A, Viani R, editors. Espresso Coffee: The Science of Quality. 2nd ed. San Diego, CA: Elsevier Academic Press; 2005; pp. 1-20. ISBN: 9780123703712-ISBN: 9780080575360

[85] Sivetz M. Growth in use of automated fluid bed roasting of coffee beans. Proceedings 14th ASIC Colloquium, Paris; 1991. pp. 313-318

[86] Wang X, Lim LT. Effect of roasting conditions on carbon dioxide degassing behavior in coffee. Food Research International. 2014;**61**:144-151. DOI: 10.1016/j.foodres.2014.01.027

[87] Moroney KM, Lee WT, O'Brien SBG, Suijver F, Marra J. Modelling of coffee extraction during brewing using multiscale methods: An experimentally validated model. Chemical Engineering Science. 2015;**137**:216-234. DOI: 10.1016/j.ces.2015.06.003

[88] Baggenstoss J, Poisson L, Kaegi R, Perren R, Escher F. Roasting and aroma formation: Effect of initial moisture content and steam treatment. Journal of Agricultural and Food Chemistry. 2008;**56**:5847-5851. DOI: 10.1021/jf8003288

[89] Ephraim D. 2016. https://www.mpechicago.com/coffee/images/uploads/pdfs/scaa_2010.pdf [Accessed: 21-02-2017]

[90] Severini C, Derossi A, Fiore AG, De Pilli T, Alessandrino O, Del Mastro A. How the variance of some extraction variables may affect the quality of espresso coffees served at coffee shop? Journal of the Science of Food and Agriculture. 2016;**96**:3023-3031. DOI: 10.1002/jsfa.7472

[91] Petracco M. Grinding. In: Illy A, Viani R, editors. Espresso Coffee: The Science of Quality. 2nd ed. San Diego, CA: Elsevier Academic Press; 2005; pp. 213-229

[92] Oliveira G, da Silva DM, Pereira RGFA, Paiva LC, Prado G, Batista LR. Effect of different roasting levels and particle sizes on ochratoxin A concentration in coffee beans. Food Control. 2013;**34**:651-656. DOI: 10.1016/j.foodcont.2013.06.014

[93] Bell LN, Wetzel CR, Grand AN. Caffeine content in coffee as influenced by grinding and brewing conditions. Food Research International. 1996;**29**:785-789. DOI: 10.1016/S0963-9969(97)00002-1

[94] Andueza S, de Peña MP, Cid C. Chemical and sensorial characteristics of espresso coffee as affected by grinding and torrefacto roast. Journal of Agricultural and Food Chemistry. 2003;**51**:7034-7039. DOI: 10.1021/jf034628f

[95] Severini C, Ricci I, Marone M, Derossi A, De Pilli T. Changes in aromatic profile of espresso coffee as a function of grinding grade and extraction time: A study by electronic nose system. Journal of Agricultural and Food Chemistry. 2015;**63**:2321-2327. DOI: 10.1021/jf505691u

[96] Severini C, Derossi A, Fiore AG, Ricci I, Marone M. The electronic nose system: Study on the global aromatic profile of espresso coffee prepared with two types of coffee filter holders. European Food Research and Technology. 2016;**242**(12):2083-2091. DOI: 10.1007/s00217-016-2705-1

[97] Spiro M, Selwood RM. The Kinetics and mechanism of caffeine infusion from coffee: The effect of particle size. Journal of the Science of Food and Agriculture. 1984;**34**:915-924. DOI: 10.1002/jsfa.2740350817

[98] Buchmann S, Kölling-Speer I, Speer K. Trigonelline and caffeine in Espresso and fresh brew coffee—Impact of grind size, water temperature, and coffee/water amount. In: The 24th International Conference on Coffee Sciences '(ASIC 2012), 12-16 November, San José, Costa Rica. ASIC Pub. - Wädenswil (Switzerland) pp. 251-255, 2012

[99] López-Galilea I, Fournier N, Cid C, Guichard E. Changes in headspace volatile concentrations of coffee brews caused by the roasting process and the brewing procedure. Journal of Agricultural and Food Chemistry. 2006;**54**:8560-8566. DOI: 10.1021/jf061178t

[100] López-Galilea I, de Peña MP, Cid C. Correlation of selected constituents with the total antioxidant capacity of coffee beverages: Influence of the brewing procedure. Journal of Agricultural and Food Chemistry. 2007;**55**:6110-6117. DOI: 10.1021/jf070779x

[101] Ludwig IA, Sanchez L, Cammerer B, Kroh LW, de Peña MP, Cid C. Extraction of coffee antioxidants: Impact of brewing time and method. Food Research International. 2012;**48**:57-64. DOI: 10.1016/j.foodres.2012.02.023

[102] Gloess AN, Schonbachler B, Klopprogge B, D'Ambrosio L, Chatelain K, Bongartz A, Strittmatter A, Rast M, Yeretzian C. Comparison of nine common coffee extraction methods: Instrumental and sensory analysis. European Food Research and Technology. 2013;**236**:607-627. DOI: 10.1007/s00217-013-1917-x

[103] Caporaso N, Genovese A, Canela MD, Civitella A, Sacchi R. Neapolitan coffee brew chemical analysis in comparison to espresso, moka and American brews. Food Research International. 2014;**61**:152-160. DOI: 10.1016/j.foodres.2014.01.020

[104] Wang X, William J, Fu Y, Lim LT. Effects of capsule parameters on coffee extraction in single-serve brewer. Food Research International. 2016;**89**:797-805. DOI:10.1016/j.foodres.2016.09.031

[105] Petracco M. Technology IV. Beverage preparation: Brewing trends for the new millennium. In: Clarke RJ and Vitzthum OG, editor. Coffee: Recent Developments. Oxford: Blackwell Science; 2001; pp. 140-164

[106] Kucukkomurler S, Özgen L. Coffee and Turkish coffee culture. Pakistan Journal of Nutrition. 2009;**8**(10):1693-1700. DOI: 10.3923/pjn.2009.1693

[107] Pérez-Martínez M, Caemmerer B, De Peña MP, Cid C, Kroh LW. Influence of brewing method and acidity regulators on the antioxidant capacity of coffee brews. Journal of Agricultural and Food Chemistry. 2010;**58**:2958-2965. DOI: 10.1021/jf9037375

[108] Santini A, Ferracane R, Mikusova P, Eged S, Srobarova A, Meca G, Manes J, Ritieni A. Influence of different coffee drink preparations on ochratoxin A content and evaluation of the antioxidant activity and caffeine variations. Food Control. 2011;**22**:1240-1245. DOI:10.1016/j.foodcont.2011.01.024

[109] Vignoli JA, Bassoli DG, Benassi MT. Antioxidant activity, polyphenols, caffeine and melanoidins in soluble coffee: The influence of processing conditions and raw material. Food Chemistry. 2011; **124**:863-868. DOI:10.1016/j.foodchem.2010.07.008

[110] Ludwig IA, Mena P, Calani L, Cid C, Del Rio D, Lean MEJ, Crozier A. Variations in caffeine and chlorogenic acid contents of coffees: What are we drinking? Food & Function. 2014;**5**:1718-1726. DOI: 10.1039/c4fo00290c

[111] Albanese D, Di Matteo M, Poiana M, Spagnamusso S. Espresso coffee (EC) by POD: Study of thermal profile during extraction process and influence of water temperature on chimica-physical and sensorial properties. Food Research International. 2009;**42**:727-732. DOI: 10.1016/j.foodres.2009.02.027

[112] Parenti A, Guerrini L, Masella P, Spinelli S, Calamai L, Spugnoli P. Comparison of espresso coffee brewing techniques. Journal of Food Engineering. 2014;**121**:112-117. DOI: 10.1016/j.foodeng.2013.08.031

[113] Derossi A, Severini C, Ricci I, Fiore AG, Caporizzi R. Espresso coffee from the single-dose capsule system. A comparison among different brands based on macroscopic and microscopic properties. Poster n.431 in The 26th International Conference on Coffee Sciences (ASIC '16); 13-20 November 2016; Kunming. Yunnan China ASIC Pub. - Wädenswil (Switzerland), 2016

[114] Barbanti D, Nicoli MC. Estrazione e stabilità della bevanda di caffè: Aspetti chimici e tecnologici. Tecnologie Alimentari. 1996;**1**:62-67

[115] Nicoli MC, Dalla Rosa M, Lerici CR. Influence of some process in conditions on solid liquid extraction of coffee. Lebensmittel-Wissenschaft & Technologie. 1990;**23**(5):386-389. DOI: 10.1006/fstl.1996.0181

[116] Andueza S, Vila AM, de Peña MP, Cid C. Influence of coffee/water ratio on the final quality of espresso coffee. Journal of the Science of Food and Agriculture. 2007;**87**:586-592. DOI: 10.1002/jsfa.2720

[117] Navarini L, Cappuccio R, Suggi-Liverani F, Illy A. Espresso coffee beverage: Classification on texture terms. Journal of Texture Studies. 2004;**35**(5):525-541. DOI: 10.1111/j.1745-4603.2004.35504.x

[118] Romani S, Severini C, Fiore AG and Pinnavaia GG. Quality of espresso coffee: A study performed through Italian coffee shops. In: International Scientific Colloquium on Coffee: Proceedings (20th ASIC); 11-15 October 2004; Bangalore, India

[119] Petracco M. Percolation. In: Illy A, Viani R, editors. Espresso Coffee: The Science of Quality. 2nd ed. San Diego, CA: Elsevier Academic Press; 2005. pp. 259-289

[120] Masella P, Guerrini L, Spinelli S, Calamai L, Spugnoli P, Illy F, Parenti A. A new espresso brewing method. Journal of Food Engineering. 2015;**146**:204-208. DOI: 10.1016/j.foodeng.2014.09.001

[121] Andueza S, Maeztu L, Dean B, de Peña MP, Bello J, Cid C. Influence of water pressure on the final quality of Arabica espresso coffee. Application of multivariate analysis. Journal of Agricultural and Food Chemistry. 2002;**50**:7426-7431. DOI: 10.1021/jf0206623

[122] Salamanca CA, Fiol N, Gonzalez C, Saez M, Villaescusa I. Extraction of espresso coffee by using gradient of temperature. Effect on physicochemical and sensorial characteristics of espresso. Food Chemistry. 2017;**214**:622-630. DOI: 10.1016/j.foodchem.2016.07.120

[123] Mora I, Rodriguez A. Comportamiento de la cafeina en un espresso. Revista del Fórum Cultural del Café. 2010;**42**:6-13

[124] Severini C, Romani S, Pinnavaia GG. Indagine sulla qualità del caffè espresso. Industrie Alimentari. 1997;**36**:12-16

[125] American Coffee Device. Available from: https://delishably.com/beverages/Coffee-Bean-Grinder [Accessed: 21-02-2017]

[126] Tfouni SAV, Carreiro LB, Teles CRA, Furlani RPZ, Cipolli KMVAB, Camargo MCR. Caffeine and chlorogenic acids intake from coffee brew: Influence of roasting degree and brewing procedure. International Journal of Food Science and Technology. 2014;**49**:747-752. DOI: 10.1111/ijfs.12361

[127] Navarini L, Nobile E, Pinto F, Scheri A, Suggi-Liverani F. Experimental investigation of steam pressure coffee extraction in a stove-top coffee maker. Applied Thermal Engineering. 2009;**29**:998-1004. DOI: 10.1016/j.applthermaleng.2008.05.014

[128] Coffee Preparation Methods. Available from: http://www.salabar.it/node/320 [Accessed: 21-02-2017]

[129] Neapolitan Coffee. Available from: http://www.mein-italien.info/getraenke/caffe.htm [Accessed: 21-02-2017]

[130] Clarke RJ. Coffee: Instant. In: Encyclopedia of Food Sciences and Nutrition. 2nd ed; 2003. pp. 1493-1498. Available from: http://lenscoffee.com [Accessed: 21-20-2017]

[131] Bartel C, Mesias M, Morales FJ. Investigation on the extractability of melanoidins in portioned espresso coffee. Food Research International. 2015;**67**:356-365. DOI: 10.1016/j.foodres.2014.11.053

[132] Caffeine Informer. Available from: https://www.caffeineinformer.com/ [Accessed: 21-02-2017]

[133] Starbucks. Available from: https://www.starbucks.com/menu/drinks/brewed-coffee/pike-place-roast#size=3 [Accessed: 21-02-2017]

Chapter 4

Caffeine Dose-Response Relationship and Behavioral Screening in Zebrafish

Luana C. Santos, Julia Ruiz-Oliveira,
Priscila F. Silva and Ana C. Luchiari

Additional information is available at the end of the chapter

http://dx.doi.org/10.5772/intechopen.68341

Abstract

It has been centuries since humans consume coffee and get the benefits of this bean. Many researches worldwide continue to show healthful properties of coffee, while others suggest a number of side effects. In fact, anything consumed in excess may cause disturbance of the body functioning, whereas caffeine is a central nervous system stimulant that increases focus and improves performance, its high concentration can cause insomnia, dizziness, and vomiting. Thus, the question is: which coffee dose promotes benefits and prevents risks? To answer it, we used the zebrafish, a popular animal model that is at the vanguard of psychopharmacological research due to its unique combination of complexity and simplicity, translational relevance and applicability to high throughput behavioral drug screens. In the current study, we examine time-course and dose-dependent changes in zebrafish following exposure to caffeine. Our data show an inverted U-shaped path for the locomotor parameters and crescent path for the anxiety-like parameters. High doses are harmful to the individual, because the stimulating effect disappears and anxiogenic effects take place. We conclude that temporal analysis of zebrafish behavior is a sensitive method for the study of acute caffeine exposure–induced functional changes in the vertebrate brain.

Keywords: *Danio rerio*, anxiety-like, drug therapy, pharmacology, biphasic effect

1. Introduction

Caffeine is a psychostimulant substance worldwide used, which can potentially increase alertness and decrease fatigue and drowsiness [1–5]. Coffee, the beverage in which caffeine is most representative, is known to be rich in biologically active compounds that possess a variety of therapeutic and functional effects. However, heavy coffee consumption may provoke systemic damages such as irregular heart rate, increased ventilation, anxiety, and due to its psychoactive properties, caffeine is likely to have addictive properties [6, 7]. Unlike other psychoactive drugs, caffeine consumption is legal and does not present any form of regulation. Moreover, caffeine consumption is not restricted to coffee and tea, but it has commonly been combined with other food products, such as chocolates, sodas, potato chips, and also bottled water. Even more risky, caffeine has been associated with other psychoactive drugs, as alcohol (i.e., energy drinks). While there is no specific recommendation on the amount of caffeine used or an indication of a critical value that may cause health problems, the U.S. Food and Drug Administration (FDA) has suggested 400 mg of caffeine/day for health adults [8]. However, there is not a clear picture of the overtime and dose-dependent effects of caffeine, demanding attention on behavioral pharmacological studies on this issue.

During the past 2 decades, numerous studies have approached the effects of psychoactive compounds used by humans in other mammals (rodents) [9–11]. Caffeine has gained attention because of its multiple targets in the brain. For instance, adenosine, ryanodine, and γ-aminobutyric acid receptors and cyclic nucleotide phosphodiesterase isoenzymes [12] seem to be related to stimulant effects of caffeine. However, as other drugs, caffeine empowers the central nervous system functioning when it reaches the therapeutic range; otherwise, it is too low concentration to cause an effect or too high concentration that causes intoxication. In this sense, several studies show contradictory behavioral effects following caffeine exposure: both increase in locomotor activity [13] and decrease in motor response [14, 15] were observed. The divergences on results may be related to caffeine dose and observation period in each study. Therefore, it is urgent to present effect of different doses and a short time-scale evaluation of caffeine induced changes in order to establish its therapeutic range, and then, how it can be properly used when in combination with other stimulant or depressant drugs.

Instead of using the most common animal model in pharmacological research, we propose the use of zebrafish to fill this gap in caffeine research. This small vertebrate is at the vanguard of neuroethological research and has been suggested for behavioral screening of drugs. The zebrafish has gained attention in behavioral brain research due to its ideal balance between the complexity of the physiological system and the simplicity of the biological model. It includes the fact that the zebrafish presents several molecular pathways, proteins, and protein products also found in mammals [16–22], besides the genome homology of about 70–80% [23]. Also, its brain structure [24, 25] and neurochemistry [26] offer translational relevance to humans [23] and allow exploring the model for a thorough understanding of the effects of substances used/abused by humans. In fact, various studies have shown that zebrafish respond similar to mammals when treated with many pharmacological compounds [27–29]. For example, benzodiazepine medication causes sedative effects in mouse [30, 31] and zebrafish [32]. For caffeine, it is not different: both rodents and zebrafish present anxiety-like behavior following caffeine exposure [33–36]. Furthermore, caffeine is water soluble and can be delivered to the fish via noninvasive method.

Zebrafish is not only an ideal model for behavioral screening, but also the majority of the genes identified in this species is conserved and has homologs in mammals [23, 37], which allows for the examination of brain function and the development of brain diseases [27]. The zebrafish is an important model for research on psychoactive substances; in this sense, to know the effect of different doses of caffeine in their behavior is an important step for the development of methodologies to assess the effect of the substance in physiology and cognition. Our overtime dose-response analyses are one of the most detailed studies of caffeine in zebrafish and serves as a behavioral screen for future studies on the neural effects of caffeine or its effects when combined with other drugs.

2. Materials and methods

2.1. Animals and housing

Adult zebrafish (wild-type, both sexes) was obtained from a local fish farm (Natal-RN) and held in 50 L tanks forming a closed recirculating high-density system at the vivarium of the Fish Laboratory (Physiology Department—UFRN). The system maintained water quality by a multistage filtration, in which four filters processed the water in a flow of 3200 L/h, including a mechanical, a biological, activated carbon, and a UV light sterilizing filter. Water temperature was maintained at 28 C, pH in 7.1. Photoperiod was set at 12:12 light:dark cycle, with light intensity of 250 lx.

Fishes were fed two times a day with live brine shrimp and flaked food. Experimental procedures were revised and approved by the Ethical Committee for Animal Use of Federal University of Rio Grande do Norte (CEUA 045/2015).

2.2. Caffeine exposure

To determine overtime effect of caffeine doses in zebrafish, 144 animals (both sexes; 4.87 ± 1.35 g) were randomly assigned to different experimental groups that corresponded to each caffeine concentration ($n = 12$ for each group). This experimental design utilized 12 acute challenge doses: 0.0 (control), 0.5, 1.0, 5.0, 10.0, 15.0, 25.0, 50.0, 65.0, 75.0, 100.0, and 150.0 mg/L caffeine (Sigma Aldrich, 1,3,7-trimethylxanthine, Cat#C0750).

Fishes were initially held in groups of 12 in glass tanks (50 cm × 30 cm × 25 cm, width × depth × height; 37 L) for 7 days to acclimatize the fish to the test room. The bottom and back side of the holding tanks were covered with white paper to provide a uniform environment. During this period, water quality was kept the same as in the stock condition, with filtration and oxygen renovation given by a 140 Bio Wheel power filters. Food was offered twice daily.

For the behavioral assay, smaller tanks (40 cm × 20 cm × 25 cm, 15 L) were used. Caffeine was added directly to the testing water, to achieve each testing dose. Fishes were individually transferred to the testing tanks and behavior was recorded during 60 min using an HD camcorder (Sony Digital Video Camera Recorder; DCR-SX45) positioned 1 m away and in front of the tanks.

Fish behavior records were tracked using the Zebtrack software developed in MatLab [38]. The behavioral variables measured were average swimming speed, total distance travelled, duration of immobility (freezing), and time spent at the bottom of the tank (up to 5 cm from the bottom).

2.3. Statistical analysis

Data were first evaluated in search for outliers, homogeneity, normality, zero trouble, collinearity, and variables independency by inferential statistics [39]. After that, a Mixed Effects Model Analysis was applied considering the behavioral response as the response variable and the time (60 min records) and caffeine dose as the explanatory variables. The repeated measures characteristic of the data (over time data sampling) required longitudinal data analysis [39].

The exploratory analysis showed abnormal distribution and over dispersion of the residuals, and thus, a glmmPQL command (MASS package [40]) was used to develop the mixed model in the R program [41]. The mixed model showed random effect factors, which was the variation in behavior between groups, fixed effect factors, which was the caffeine doses used, and error.

The response variables were positive continuous quantitative data, not including zero ($Y > 0$); thus, a Goodness-of-fittest was performed to verify the best distribution function. If the response variable is continuous, then the normal and gamma are the best options [39]. Our response variable data best distribution function fitted gamma distribution (link function = inverse). The post-hoc comparisons between treatments, of each model, were made using the Tukey test in "lsmeans" package [42].

Average speed, distance traveled, freezing, and time at the bottom of the tank were also compared between the groups (caffeine doses) using One-Way ANOVA. For all comparison, the probability level considered for significance was $p < 0.05$.

3. Results

Figure 1 depicts zebrafish average swimming speed during acute caffeine exposure. Graphs referent to the very small doses (0.5 and 1.0 mg/L) are not presented to make it clear and due to its similarity to the doses below (0 mg/L) and above (5 mg/L). The mixed model comparison showed that average speed changed over time (GLMM, $\chi^2 = 7.33$, df = 1, $p < 0.006$; **Table 1**) and due to the caffeine dose used (GLMM, $\chi^2 = 651.49$, df = 10, $p < 0.001$; **Table 1**). The post-hoc comparison test (lsmeans) between caffeine doses is shown in **Table 1**: 10 and 25 mg/L of caffeine lead to an increase in average speed, while higher doses (50–150 mg/L) cause a decrease in swimming speed compared to low doses (0–5 mg/L).

Figure 2 shows zebrafish total distance traveled along the 60-min period of caffeine exposure. Graphs denoting the behavior for 0.5 and 1.0 m/L are not shown. The mixed model

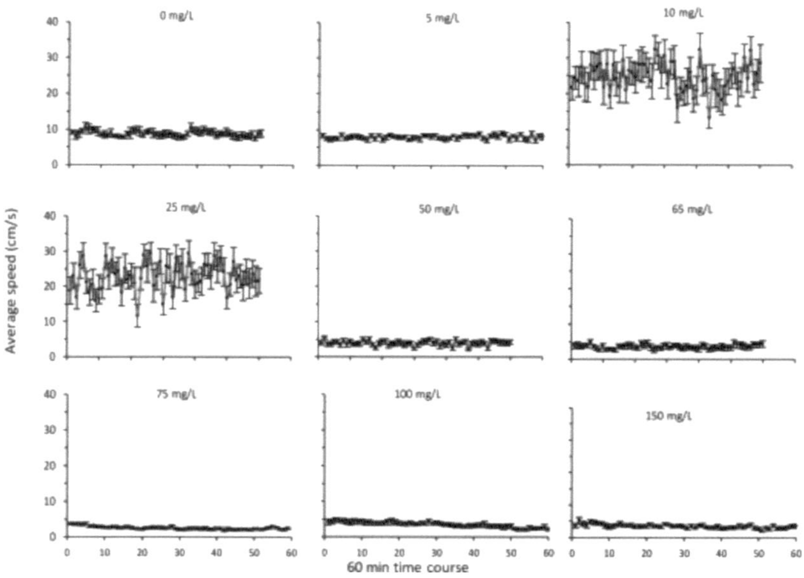

Figure 1. Time-course path of the average swimming speed during 60-min caffeine exposure in zebrafish. Mean ± SEM are shown for every 1-min intervals of the total 60-min recording. The caffeine doses (group designations) are shown above the graphs. Sample sizes (n) were 12 for each dose. Note the elevated activity in the group of fish exposed to 10 and 25 mg/L caffeine as compared to control. Also note the decreased activity in the fish that was exposed to doses of 50 mg/L caffeine and above it. For statistical analysis see Section 3 and **Table 1**.

comparison showed that total distance traveled changed over time (GLMM, $\chi^2 = 11.68$, df = 1, $p < 0.0006$; **Table 1**) and due to the caffeine dose used (GLMM, $\chi^2 = 271.49$, df = 10, $p < 0.001$; **Table 1**). The post-hoc comparison test (lsmeans) between caffeine doses is shown in **Table 1**. Caffeine doses higher than 50 mg/L decreased distance traveled compared to doses from 0 to 25 mg/L.

Figure 3 illustrates freezing behavior presented by zebrafish during caffeine challenge, a behavior related to fear and anxiety. Graphs referent to doses 0.5 and 1.0 mg/L are omitted. The mixed model comparison showed that freezing did not change over time (GLMM, $\chi^2 = 2.13$, df = 1, $p < 0.14$; **Table 1**) but changed according to the caffeine dose used (GLMM, $\chi^2 = 214.66$, df = 10, $p < 0.001$; **Table 1**). The post-hoc comparison test (lsmeans) between caffeine doses is shown in **Table 1**. Caffeine doses higher than 50 mg/L increased freezing behavior in zebrafish compared to doses from 0 to 25 mg/L.

Figure 4 presents zebrafish time spent at the bottom of the testing tank, another behavior associated to fear and anxiety response. Again, graphs representing doses 0.1 and 1.0 mg/L were not presented. The mixed model comparison showed that distance from the bottom did not change over time (GLMM, $\chi^2 = 0.18$, df = 1, $p < 0.66$; **Table 1**) but changed due to the caffeine exposure (GLMM, $\chi^2 = 170.91$, df = 10, $p < 0.001$; **Table 1**). The post-hoc comparison test (lsmeans) between caffeine doses is shown in **Table 1**. Caffeine doses of 50 mg/L and higher increased the time fish spent at the bottom of the tank.

	Behavioral parameters											
	Average speed			Total distance travelled			Freezing			Distance from bottom		
Explanatory variable	**Chi-squared**	***p*-Value**		**Chi-squared**	***p*-Value**		**Chi-squared**	***p*-Value**		**Chi-squared**	***p*-Value**	
60 min time course	7.33	0.006		11.68	0.0006		2.12	0.14		0.18	0.66	
Doses	651.49	>0.001		271.49	<0.001		214.66	<0.001		170.91	<0.001	
Pairwise comparison	**lsmeans** ± SEM	***t*-Value**	***p*-Value**	**lsmeans** ± SEM	***t*-Value**	***p*-Value**	**lsmeans** ± SEM	***t*-value**	***p*-value**	**lsmeans** ± SEM	***t*-value**	***p*-value**
Dose 0.0 vs. dose 0.5	−0.00 ± 0.01	−0.22	1.0	−4.17 ± 0.00	−0.83	0.99	0.01 ± 0.02	0.73	0.99	−0.01 ± 0.01	−0.84	0.99
Dose 0.0 vs. dose 1.0	−0.00 ± 0.01	−0.28	1.0	−3.11 ± 0.00	−0.62	0.99	0.01 ± 0.02	0.42	1.00	−0.01 ± 0.01	−0.80	0.99
Dose 0.0 vs. dose 5.0	−0.00 ± 0.01	−0.40	1.0	−3.45 ± 0.00	−0.68	0.99	0.03 ± 0.02	1.34	0.95	−0.00 ± 0.01	−0.33	1.00
Dose 0.0 vs. dose 10	0.06 ± 0.01	4.52	0.00	7.32 ± 0.00	1.46	0.92	−0.06 ± 0.02	−2.45	0.34	−0.00 ± 0.01	−0.64	0.99
Dose 0.0 vs. dose 25	0.06 ± 0.01	4.29	0.00	4.17 ± 0.00	0.83	0.99	−0.06 ± 0.02	−2.43	0.45	−0.00 ± 0.01	−0.07	1.00
Dose 0.0 vs. dose 50	−0.09 ± 0.01	−0.55	<0.00	−3.08 ± 0.00	−6.03	<0.00	0.14 ± 0.02	5.34	<0.00	0.04 ± 0.01	3.38	0.03
Dose 0.0 vs. dose 65	−0.13 ± 0.02	−8.07	<0.00	−3.74 ± 0.00	−6.69	<0.00	0.14 ± 0.02	5.25	<0.00	0.06 ± 0.01	4.50	0.00
Dose 0.0 vs. dose 75	−0.18 ± 0.01	−12.26	<0.00	−4.78 ± 0.00	−9.19	<0.00	0.15 ± 0.02	5.87	<0.00	0.08 ± 0.01	6.34	<0.00
Dose 0.0 vs. dose 100	−0.13 ± 0.01	−9.20	<0.00	−3.29 ± 0.00	−6.42	<0.00	0.07 ± 0.02	5.80	0.01	0.06 ± 0.01	5.08	0.00
Dose 0.0 vs. dose 150	−0.12 ± 0.01	−8.18	<0.00	−2.64 ± 0.00	−5.18	<0.00	0.17 ± 0.02	6.72	<0.00	0.07 ± 0.01	5.32	<0.00
Dose 0.5 vs. dose 1.0	−0.00 ± 0.01	−0.06	1.0	1.05 ± 0.00	0.20	1.00	−0.00 ± 0.02	0.30	1.00	0.00 ± 0.01	0.03	1.00

	Behavioral parameters											
	Average speed			Total distance travelled			Freezing			Distance from bottom		
Dose 0.5 vs. dose 5.0	−0.00 ± 0.01	−0.18	1.0	7.18 ± 0.00	0.14	1.00	0.01 ± 0.02	0.61	0.99	0.00 ± 0.01	0.50	1.00
Dose 0.5 vs. dose 10	0.07 ± 0.01	4.74	0.00	1.14 ± 0.00	2.29	0.44	−0.08 ± 0.02	−3.18	0.06	0.00 ± 0.01	0.19	1.00
Dose 0.5 vs. dose 25	0.06 ± 0.01	4.51	0.00	8.35 ± 0.00	1.66	0.85	−0.08 ± 0.02	−3.17	0.06	0.01 ± 0.00	0.76	0.99
Dose 0.5 vs. dose 50	−0.09 ± 0.01	−6.33	<0.00	−2.67 ± 0.00	−5.21	<0.00	0.12 ± 0.02	4.61	0.00	0.05 ± 0.01	4.23	0.00
Dose 0.5 vs. dose 65	−0.12 ± 0.02	−7.88	<0.00	−4.36 ± 0.01	−8.38	<0.00	0.12 ± 0.02	4.57	0.00	0.07 ± 0.01	5.29	<0.00
Dose 0.5 vs. dose 75	−0.18 ± 0.01	−12.04	<0.00	−2.87 ± 0.01	−5.60	<0.00	0.13 ± 0.02	5.13	0.00	0.09 ± 0.01	7.19	<0.00
Dose 0.5 vs. dose 100	−0.13 ± 0.01	−8.98	<0.00	−2.22 ± 0.01	−4.35	0.00	0.05 ± 0.02	2.07	0.60	0.08 ± 0.01	5.93	<0.00
Dose 0.5 vs. dose 150	−0.12 ± 0.01	−7.96	<0.00	−3.35 ± 0.01	−0.06	0.00	0.15 ± 0.02	5.99	<0.00	0.08 ± 0.01	6.17	<0.00
Dose 1.0 vs. dose 5.0	−0.00 ± 0.01	−0.12	1.0	1.04 ± 0.01	2.08	0.59	0.02 ± 0.01	0.92	0.99	0.00 ± 0.01	0.46	1.00
Dose 1.0 vs. dose 10	0.07 ± 0.01	4.80	0.00	7.29 ± 0.01	1.45	0.93	−0.07 ± 0.01	−2.87	0.14	0.00 ± 0.01	0.16	1.00
Dose 1.0 vs. dose 25	0.06 ± 0.01	4.57	0.00	−2.77 ± 0.01	−5.42	<0.00	−0.07 ± 0.01	−2.06	0.14	0.01 ± 0.01	0.72	0.99
Dose 1.0 vs. dose 50	−0.09 ± 0.01	−6.27	<0.00	−3.43 ± 0.01	−6.12	<0.00	0.12 ± 0.02	4.92	0.00	0.05 ± 0.01	4.19	0.00
Dose 1.0 vs. dose 65	−0.12 ± 0.02	−7.82	<0.00	−4.47 ± 0.01	−8.58	<0.00	0.13 ± 0.02	4.85	0.00	0.07 ± 0.01	5.26	<0.00
Dose 1.0 vs. dose 75	−0.18 ± 0.01	−11.98	<0.00	−2.98 ± 0.00	−5.80	<0.00	0.14 ± 0.01	5.44	<0.00	0.09 ± 0.01	7.16	<0.00
Dose 1.0 vs. dose 100	−0.13 ± 0.01	−8.92	<0.00	−2.33 ± 0.00	−4.56	0.00	0.06 ± 0.02	2.38	0.38	0.08 ± 0.01	5.90	<0.00

	Behavioral parameters											
	Average speed			Total distance travelled			Freezing			Distance from bottom		
Dose 1.0 vs. dose 150	−0.11 ± 0.01	−7.90	<0.00	1.07 ± 0.00	2.15	0.04	0.16 ± 0.01	6.30	<0.00	0.08 ± 0.01	6.14	<0.00
Dose 5.0 vs. dose 10	0.07 ± 0.01	4.93	0.00	1.07 ± 0.00	2.15	0.54	−0.10 ± 0.00	−3.80	0.01	−0.00 ± 0.01	−0.30	1.00
Dose 5.0 vs. dose 25	0.06 ± 0.01	4.69	0.00	7.63 ± 0.00	1.52	0.90	−0.09 ± 0.01	−3.78	0.01	0.00 ± 0.01	0.25	1.00
Dose 5.0 vs. dose 50	−0.09 ± 0.01	−6.14	<0.00	−2.74 ± 0.00	−5.35	<0.00	0.10 ± 0.01	4.00	0.00	0.05 ± 0.01	3.73	0.01
Dose 5.0 vs. dose 65	−0.12 ± 0.02	−7.71	<0.00	−3.39 ± 0.00	−6.06	<0.00	0.11 ± 0.02	4.00	0.00	0.07 ± 0.01	4.83	0.00
Dose 5.0 vs. dose 75	−0.18 ± 0.01	−11.86	<0.00	−4.43 ± 0.00	−8.52	<0.00	0.11 ± 0.02	4.52	0.00	0.09 ± 0.01	6.70	<0.00
Dose 5.0 vs. dose 100	−0.13 ± 0.01	−8.80	<0.00	−2.94 ± 0.00	−5.74	<0.00	0.03 ± 0.00	1.45	0.93	0.07 ± 0.01	5.43	<0.00
Dose 5.0 vs. dose 150	−0.11 ± 0.01	−7.78	<0.00	−2.29 ± 0.01	−4.49	0.00	0.14 ± 0.02	5.38	<0.00	0.07 ± 0.01	5.68	<0.00
Dose 10 vs. dose 25	0.00 ± 0.01	−0.23	1.00	−3.14 ± 0.00	−0.63	0.99	0.00 ± 0.00	0.01	1.00	0.00 ± 0.01	0.56	1.00
Dose 10 vs. dose 50	−0.16 ± 0.01	−11.04	<0.00	−3.82 ± 0.00	−7.48	<0.00	0.20 ± 0.02	7.79	<0.00	0.05 ± 0.01	4.03	0.00
Dose 10 vs. dose 65	−0.19 ± 0.02	−12.19	<0.00	−4.47 ± 0.00	−8.01	<0.00	0.21 ± 0.02	7.52	<0.00	0.07 ± 0.01	5.11	0.00
Dose 10 vs. dose 75	−0.25 ± 0.01	−16.68	<0.00	−5.51 ± 0.01	−10.62	<0.00	0.21 ± 0.02	8.32	<0.00	0.09 ± 0.01	7.00	<0.00
Dose 10 vs. dose 100	−0.20 ± 0.01	−13.66	<0.00	−4.02 ± 0.00	−7.87	<0.00	0.13 ± 0.00	5.25	<0.00	0.07 ± 0.01	5.74	<0.00
Dose 10 vs. dose 150	−0.19 ± 0.01	−12.66	<0.00	−3.37 ± 0.00	−6.63	<0.00	0.24 ± 0.02	9.17	<0.00	0.08 ± 0.01	5.98	<0.00
Dose 25 vs. dose 50	−0.16 ± 0.01	−10.81	<0.00	−3.50 ± 0.01	−6.86	<0.00	0.20 ± 0.02	7.78	<0.00	0.04 ± 0.01	3.47	0.02

http://dx.doi.org/10.5772/intechopen.68341

	Behavioral parameters											
	Average speed			Total distance travelled			Freezing			Distance from bottom		
Dose 25 vs. dose 65	−0.19 ± 0.02	−11.98	<0.00	−4.16 ± 0.00	−7.44	<0.00	0.21 ± 0.02	7.50	<0.00	0.06 ± 0.01	4.59	0.00
Dose 25 vs. dose 75	−0.25 ± 0.01	−16.45	<0.00	−5.20 ± 0.01	−10.00	<0.00	0.21 ± 0.00	8.30	<0.00	0.08 ± 0.01	6.44	<0.00
Dose 25 vs. dose 100	−0.20 ± 0.01	−13.43	<0.00	−3.71 ± 0.00	−7.24	<0.00	0.13 ± 0.01	5.24	<0.00	0.07 ± 0.01	5.18	<0.00
Dose 25 vs. dose 150	−0.18 ± 0.01	−12.42	<0.00	−3.05 ± 0.00	−6.00	<0.00	0.24 ± 0.01	9.16	<0.00	0.17 ± 0.01	5.42	<0.00
Dose 50 vs. dose 65	−0.03 ± 0.02	−2.05	0.60	−6.55 ± 0.00	−1.15	0.98	0.00 ± 0.02	0.29	1.00	0.01 ± 0.01	1.35	0.95
Dose 50 vs. dose 75	−0.08 ± 0.01	−5.76	0.08	−1.69 ± 0.00	−3.19	0.06	0.01 ± 0.00	0.52	1.00	0.03 ± 0.01	2.98	0.11
Dose 50 vs. dose 100	−0.04 ± 0.01	−2.67	0.22	−2.05 ± 0.00	−0.39	1.00	−0.06 ± 0.02	−2.54	0.29	0.02 ± 0.01	1.70	0.82
Dose 50 vs. dose 150	−0.02 ± 0.01	−1.64	0.85	4.47 ± 0.01	0.86	0.99	0.03 ± 0.02	1.38	0.95	0.02 ± 0.01	1.95	0.67
Dose 65 vs. dose 75	−0.05 ± 0.02	−3.28	0.05	−1.03 ± 0.01	−1.80	0.77	0.00 ± 0.02	0.18	1.00	0.02 ± 0.01	1.40	0.94
Dose 65 vs. dose 100	−0.00 ± 0.02	−0.41	1.00	4.50 ± 0.00	0.79	0.99	−0.07 ± 0.02	−2.65	0.23	0.00 ± 0.01	0.22	1.00
Dose 65 vs. dose 150	0.00 ± 0.02	0.53	1.00	1.10 ± 0.01	1.94	0.68	0.02 ± 0.02	0.98	0.99	0.00 ± 0.01	0.45	1.00
Dose 75 vs. dose 100	0.04 ± 0.01	3.09	0.08	1.48 ± 0.00	2.80	0.16	−0.08 ± 0.01	−3.06	0.09	−0.01 ± 0.01	−1.28	0.97
Dose 75 vs. dose 150	0.06 ± 0.01	4.12	0.07	2.14 ± 0.00	4.05	0.07	0.02 ± 0.02	0.85	0.99	−0.01 ± 0.01	−1.02	0.99
Dose 100 vs. dose 150	0.01 ± 0.01	1.03	0.99	6.52 ± 0.00	1.25	0.97	0.10 ± 0.02	3.92	0.00	0.00 ± 0.01	0.25	1.00

Dose values are expressed in mg/L.

Table 1. Estimates of mixed effect model for the behavioral parameters measured during caffeine exposure.

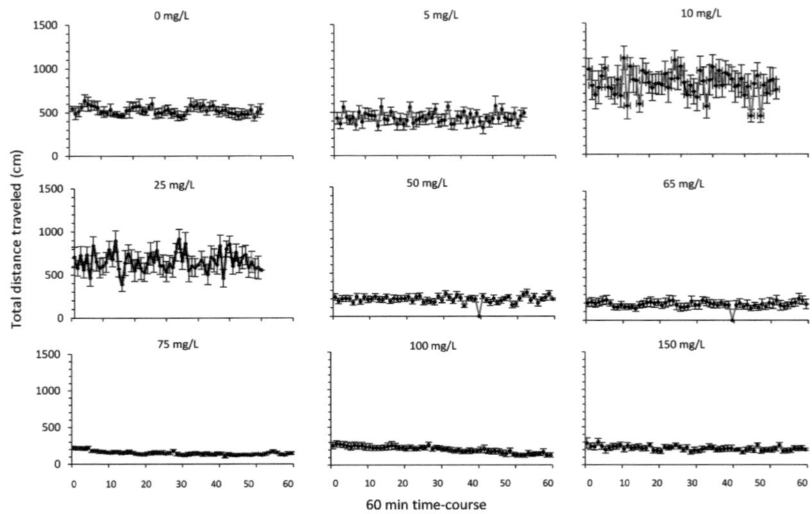

Figure 2. Time-course path of the total distance traveled during 60-min caffeine exposure in zebrafish. Mean ± SEM are shown for every 1-min intervals of the total 60 min recording. The caffeine doses (group designations) are shown above the graphs. Sample sizes (*n*) were 12 for each dose. Note the elevated activity in the group of fish exposed to 10 and 25 mg/L caffeine as compared to control. Also note the decreased activity in the fish that was exposed to doses of 50 mg/L caffeine and above it. For statistical analysis see Section 3 and **Table 1**.

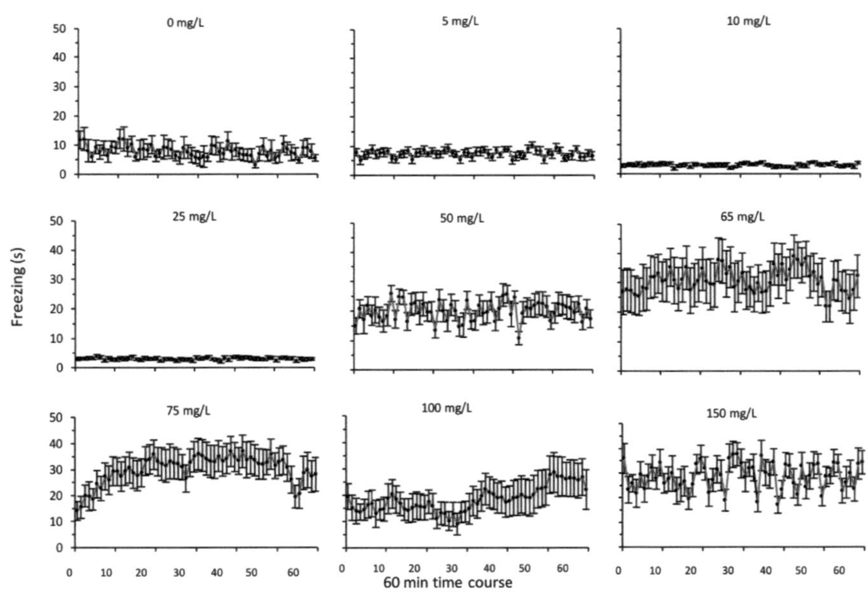

Figure 3. Time-course path of the freezing behavior during 60-min caffeine exposure in zebrafish. Mean ± SEM are shown for every 1-min intervals of the total 60 min recording. The caffeine doses (group designations) are shown above the graphs. Sample sizes (*n*) were 12 for each dose. Note the decreased freezing in the group of fish exposed to 5, 10 and 25 mg/L caffeine as compared to control. On the contrary, note the increased freezing in the fish that was exposed to doses above 50 mg/L caffeine. For statistical analysis, see Section 3 and **Table 1**.

http://dx.doi.org/10.5772/intechopen.68341

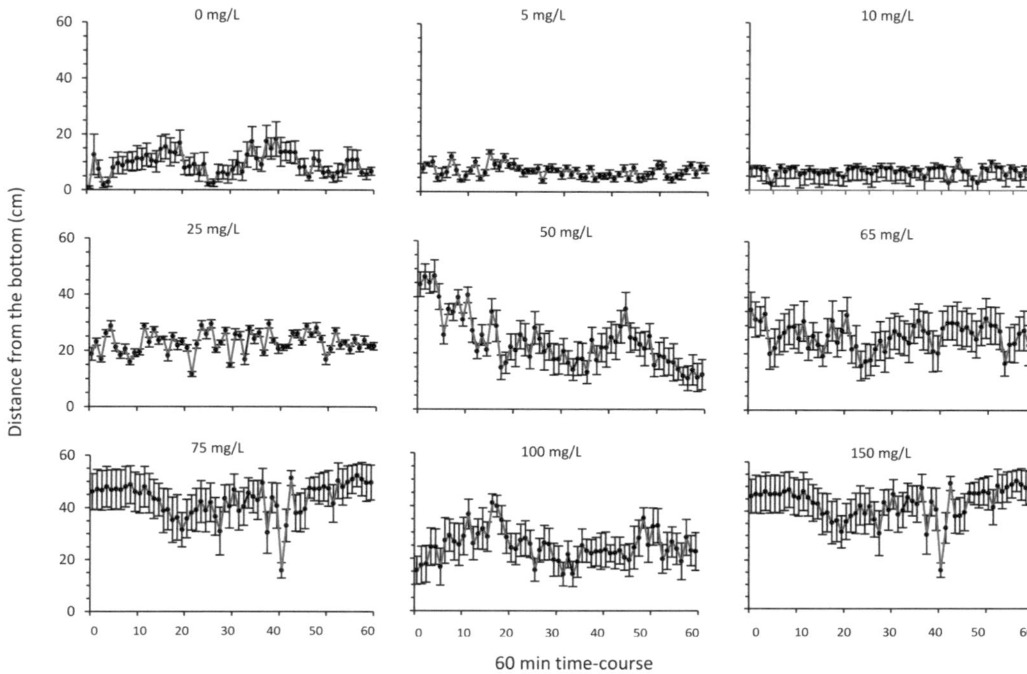

Figure 4. Time-course path of the time spent at the bottom of the tank (up to 5 cm from the bottom) during 60-min caffeine exposure in zebrafish. Mean ± SEM are shown for every 1-min intervals of the total 60 min recording. The caffeine doses (group designations) are shown above the graphs. Sample sizes (*n*) were 12 for each dose. Note the decreased time at the bottom in the group of fish exposed to 5, 10, and 25 mg/L caffeine as compared to control. Also note the increased time at the bottom in the fish exposed to doses above 50 mg/L caffeine. For statistical analysis, see Section 3 and **Table 1**.

Figure 5 displays the median and range of variation of the 60-min caffeine exposure in zebrafish, both for the parameters related to locomotion (**Figure 5a** and **b**) and the parameters related to anxiety-like behavior (**Figure 5c** and **d**). One-Way ANOVA between groups (caffeine doses) showed that average speed did not differ from the control condition when fish is exposed to doses up to 5 mg/L caffeine, but it is increased with doses of 10 and 25 mg/L and decreased with doses above 50 mg/L ($F = 1087.97$, df = 10, $p < 0.001$, **Figure 5a**) indicating an inverted U shape. The same patterns were observed for total distance travelled, in which the lower doses (0.5–5 mg/L) did not differ from the control, doses of 10 and 25 mg/L increased distance traveled and doses above 50 mg/L decreased distance traveled (One-Way ANOVA, $F = 374.82$, df = 10, $p < 0.001$, **Figure 5b**). For the freezing behavior, One-Way ANOVA showed a slight decrease with increasing doses, with the lowest values of freezing registered at 10 and 25 mg/L, and a sharp increase with doses above 50 mg/L ($F = 462.15$, df = 10, $P < 0.001$, **Figure 5c**), suggesting the anxiogenic effect of high caffeine doses. The time fish spent at the bottom of the tank was reduced by doses from 0.5 to 25 mg/L and highly increased by doses above 50 mg/L (One-Way ANOVA, $F = 427.27$, df = 10, $p < 0.001$). **Figure 5d** depicts the comparison between caffeine doses and the tendency line indicating an increasing in time at the bottom concomitant to the increase in caffeine dose.

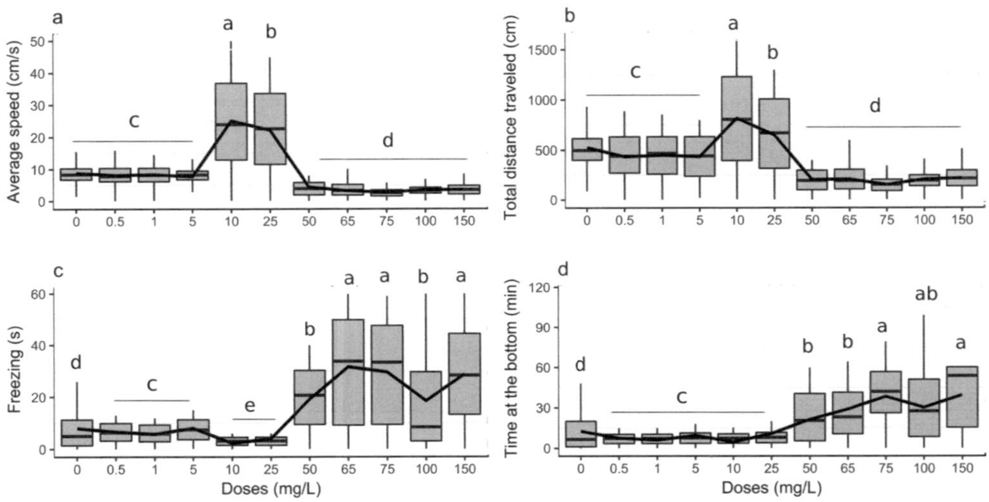

Figure 5. Box plot shows median and interquartile range of locomotor behavior: (a) average swimming speed, (b) total distance traveled, (c) freezing, and (d) time spent at the bottom of the tank, whiskers represent the range. Tendency line indicates the inverted U shape observed for increasing doses of caffeine on locomotor behavioral parameters (swimming speed and distance traveled), and the ascendant pattern observed for increase doses of caffeine on anxiety-like behavioral parameters (freezing and time at the bottom of the tank). For statistical analysis, see Section 3.

4. Discussion

This study characterizes variations in the locomotor pattern and anxiety-like behavior derived from acute exposure to caffeine in zebrafish. We evaluated a wide spectrum of caffeine doses, from 0.5 to 150 mg/L, every 1-min interval along 60-min period of the drug exposure. This detailed analysis of different doses overtime allowed to the observation that caffeine has a biphasic effect, stimulating locomotion and decreasing anxiety at low levels, and diminishing activity and increasing anxiety-like behavior at doses above 50 mg/L.

While caffeine is accepted to act as a stimulant on the central nervous system [43], it is worth noting that in fact this drug exerts distinct responses depending on the amount used. Initially, caffeine has little or no impact on behavior, followed by the most evident effect caused by an intermediate dose, and then a remarkable suppression at the level of behavioral activity, as observed in **Figure 5**. High caffeine doses (50 and 100 mg/L) were previously shown to depress locomotor activity in zebrafish [15, 44], but our study is the first to present a time-course analyses of the effects of several doses of caffeine in adult zebrafish. The same dose-dependent effects observed herein in rodents, suggesting the models similarity in terms of the mechanism by which caffeine produces behavioral effects [45].

Adenosine receptor blockade seems to be the prevalent action of caffeine in the brain. The increase on locomotor activity after caffeine exposure derives from the blockade of A_{2A} adenosine receptors, preventing the inhibitory action of adenosine on the nervous system [46]. The antagonist role of caffeine usually stimulates the central nervous system and also activates dopaminergic transmission [47–49], which is consistent with the drug reinforcing properties. Another event that occurs at the time of caffeine ingestion and which causes increased locomotor activity is the inhibition of phosphodiesterase (an enzyme that hydrolyses cAMP), which promotes the release of calcium from intracellular reserves and interferes with the sensitivity of GABA receptors [50].

Caffeine is a substance widely used by the society [1] and, if used in moderation (up to 200 mg/day/person on average), it may lead to several benefits, such as improved performance on tasks that require attention and focus [51, 52]. For instance, it was observed that zebrafish improves object discrimination when treated with moderate doses of caffeine [53]. The most notable effects of low-to-moderate doses of caffeine include increased alertness, energy, and ability to concentrate [54]. On the other hand, high and abusive use of caffeine may inhibit these effects [55]. At high concentrations, caffeine is suggested to increase glucose utilization in the CNS, what also seems to be related to its stimulatory effects [56]. The elevated sugar level (main CSN subtract) together with the blockage of adenosine inhibitory effects, in turn, is responsible to the caffeine side-effects on the motor system and sleep-wake cycle, two functions highly susceptible to caffeine.

Moreover, higher caffeine doses induce negative effects such as increased sympathetic response (tachycardia, higher ventilation), restlessness, insomnia, and anxiety [57]. The high acute dose of caffeine exacerbates anxiety-like behavior, reduces locomotion, and in many cases, causes behavior similar to seizure [58], very high doses of caffeine may also cause intoxication and death of the individuals [54, 59]. The same pattern of effect was observed in zebrafish larvae under the action of high doses of caffeine [60]. It is known that adenosine regulates the activity of several neurons, such as glutamatergic; thus, if the effect of adenosine is blocked, glutamatergic transmission increases and may turn to an extremely high excitatory response [12, 56].

Finally, our results reinforce the zebrafish as a valuable model organism for throughput screening of behavioral-related drugs. While caffeine is a legal and widely consumed substance, the amount of caffeine ingested should be taken into consideration since negative effects are observed after high doses consumption. We found that moderate caffeine intake ameliorates performance, but a robust anxiety-related response occurs following exposure to high doses of caffeine. Taken together, these results confirm zebrafish as an accurate, reliable, and efficient model for basic translational research of psychoactive drugs on physiology and behavior.

Acknowledgements

The authors are thankful to J.J. Oliveira for technical support and F. Freire for statistical assistance.

Author details

Luana C. Santos, Julia Ruiz-Oliveira, Priscila F. Silva and Ana C. Luchiari*

*Address all correspondence to: analuchiari@yahoo.com.br

Department of Physiology, Bioscience Center, Federal University of Rio Grande do Norte, Natal, Rio Grande do Norte, Brazil

References

[1] Fredholm BB, Bättig K, Holmén J, Nehlig A, Zvartau EE. Actions of caffeine in the brain with special reference to factors that contribute to its widespread use. Journal of Pharmacology and Experimental Therapeutics. 1999;**51**(1):83–133. DOI: 0031-6997/99/5101-0083$03.00/0

[2] Frary CD, Johnson RK, Wang MQ. Food sources and intakes of caffeine in the diets of persons in the United States. Journal of the American Dietetic Association. 2005;**105**(1):110–113. DOI: 10.1016/j.jada.2004.10.027

[3] Mitchell DC, Knight CA, Hockenberry J, Teplansky R, Hartman TJ. Beverage caffeine intakes in the U.S. Food and Chemical Toxicology. 2014;**63**:136–142. DOI: 10.1016/j.fct.2013.10.042

[4] Heckman MA, Weil J, Mejia EG. Caffeine (1,3,7-trimethylxanthine) in foods: A comprehensive review on consumption, functionality, safety, and regulatory matters. Journal of Food Science. 2010;**75**(3):77–87. DOI: 10.1111/j.1750-3841.2010.01561.x

[5] Beydoun MA, Gamaldo AA, Beydoun HA, Tanaka T, Tucker KL, Talegawkar SA, Ferrucci L, Zonderman AB. Caffeine and alcohol intakes and overall nutrient adequacy are associated with longitudinal cognitive performance among U.S. adults. The Journal of Nutrition. 2014;**144**(6):890–901. DOI: 10.3945/jn.113.189027

[6] Budygin EA, Weiner JL. Exploring the neurochemical basis of alcohol addiction-related behaviors: Translational research. Translational Biomedicine. 2015;**6**(Suppl Spec). https://www.ncbi.nlm.nih.gov/pmc/articles/PMC4710378/pdf/nihms723979.pdf

[7] Ferré S. Mechanisms of the psychostimulant effects of caffeine: Implications for substance use disorders. Psychopharmacology. 2016;**233**(10):1963–1979. DOI: 10.1007/s00213-016-4212-2

[8] U.S. Food and Drug Administration. Citizen Petition for New Labeling Requirement for Caffeine in Food [Internet]. 2016. Available from: http://www.fda.gov/ohrms/dockets/dockets/06p0039/06p-0039-cp00001-01-vol1.pdf [Accessed: 05-01-2017]

[9] Molloy JW, Calcagno CJ, Williams CD, Jones FJ, Torres DM, Harrison SA. Association of coffee and caffeine consumption with fatty liver disease, nonalcoholic steatohepatitis, and degree of hepatic fibrosis. Hepatology. 2012;**55**(2):429–446. DOI: 10.1002/hep.24731

[10] Sepkowitz KA. Energy drinks and caffeine-related adverse effects energy drinks and caffeine-related adverse effects. JAMA. 2013;**309**(3):243–244. DOI: 10.1001/jama.2012.173526

[11] Denoeud F, Carretero-Paulet L, Dereeper A, Droc G, Guyot R, Pietrella M, Zheng C, Alberti A, Anthony F, Aprea G, Aury JM, Bento P, Bernard M, Bocs S, Campa C, Cenci A, Combes MC, Crouzillat D, Da Silva C, Daddiego L, De Bellis F, Dussert S, Garsmeur O, Gayraud T, Guignon V, Jahn K, Jamilloux V, Joët T, Labadie K, Lan T, Leclercq J, Lepelley M, Leroy T, Li LT, Librado P, Lopez L, Muñoz A, Noel B, Pallavicini A, Perrotta G, Poncet V, Pot D,

Priyono, Rigoreau M, Rouard M, Rozas J, Tranchant-Dubreuil C, VanBuren R, Zhang Q, Andrade AC, Argout X, Bertrand B, de Kochko A, Graziosi G, Henry RJ, Jayarama, Ming R, Nagai C, Rounsley S, Sankoff D, Giuliano G, Albert VA, Wincker P, Lashermes P. The coffee genome provides insight into the convergent evolution of caffeine biosynthesis. Science. 2014;**345**(6201):1181–1184. DOI: 10.1126/science.1255274

[12] Fisone G, Borgkvist A, Usiello A. Caffeine as a psychomotor stimulant: Mechanism of action. Cellular and Molecular Life Sciences. 2004;**61**(7–8):857–872. DOI: 10.1007/s00018-003-3269-3

[13] Maximino C, da Silva AW, Gouveia AJ, Herculano AM. Pharmacological analysis of zebrafish (*Danio rerio*) scototaxis. Progress in Neuro-Psychopharmacology & Biological Psychiatry. 2011;35(2):624–631. DOI: 10.1016/j.pnpbp.2011.01.006

[14] Pechlivanova D, Tchekalarova J, Nikolov R, Yakimova K. Dose-dependent effects of caffeine on behavior and thermoregulation in a chronic unpredictable stress model of depression in rats. Behavioural Brain Research. 2010;**209**(2):205–211. DOI: 10.1016/j.bbr.2010.01.037

[15] Ladu F, Mwaffo V, Li J, Macrì S, Porfiri M. Acute caffeine administration affects zebrafish response to a robotic stimulus. Behavioural Brain Research. 2015;**289**:48–54. DOI: 10.1016/j.bbr.2015.04.020

[16] Faraco JH, Appelbaum L, Marin W, Gaus SE, Mourrain P, Mignot E. Regulation of hypocretin (orexin) expression in embryonic zebrafish. Journal of Biological Chemistry. 2006;**281**(40):29753–29761. DOI: 10.1074/jbc.M605811200

[17] Holzschuh J, Ryu S, Aberger F, Driever W. Dopamine transporter expression distinguishes dopaminergic neurons from other catecholaminergic neurons in the developing zebrafish embryo. Mechanisms of Development. 2001;**101**(1–2):237–243. DOI: 10.1016/S0925-4773(01)00287-8

[18] Kaslin J, Nystedt JM, Ostergård M, Peitsaro N, Panula P. Journal of Neuroscience. 2004;**24**(11):2678–2689. DOI: 10.1523/JNEUROSCI.4908-03.2004

[19] Kaslin J, Panula P. Comparative Anatomy of the Histaminergic and Other Aminergic Systems in zebrafish (*Danio rerio*). Journal of Comparative Neurology. 2001;**440**(4):342–377. DOI: 10.1002/cne.1390

[20] McLean DL, Fetcho JR. Ontogeny and innervation patterns of dopaminergic, noradrenergic, and serotonergic neurons in larval zebrafish. Journal of Comparative Neurology. 2004;**480**(1):38–56. DOI: 10.1002/cne.20280

[21] Mueller T, Vernier P, Wullimann MF. The adult central nervous cholinergic system of a neurogenetic model animal, the zebrafish *Danio rerio*. Brain Research. 2004;**1011**(2):156–169.

[22] Prober DA, Rihel J, Onah AA, Sung RJ, Schier AF. Hypocretin/orexin overexpression induces an insomnia-like phenotype in Zebrafish. Journal of Neuroscience. 2006;**26**(51):13400–13410. DOI: 10.1523/JNEUROSCI.4332-06.2006

[23] Crollius HR, Weissenbach J. Fish genomics and biology. Genome Research. 2005;**15**(12):1675–1682. DOI: http://www.genome.org/cgi/doi/10.1101/gr.3735805

[24] Klee EW, Schneider H, Clark KJ, Cousin MA, Ebbert JO, Hooten WM, Karpyak VM, Warner DO, Ekker SC. Zebrafish: A model for the study of addiction genetics. Human Genetics. 2012;**131**(6):977–1008. DOI: 10.1007/s00439-011-1128-0

[25] Kolb B, Whishaw IQ. Brain plasticity and behavior. Annual Review of Psychology. 1998;**49**(1):43–64. DOI: 10.1146/annurev.psych.49.1.43

[26] Gerlai R, Chatterjee D, Pereira T, Sawashima T, Krishnannair R. Acute and chronic alcohol dose: Population differences in behavior and neurochemistry of zebrafish. Genes, Brain, and Behavior. 2009;**8**(6):586–599. DOI: 10.1111/j.1601-183X.2009.00488.x

[27] Kalueff AV, Stewart AM, Gerlai R. Zebrafish as an emerging model for studying complex brain disorders. Trends in Pharmacological Sciences. 2014;**35**(2):63–75. DOI: 10.1016/j.tips.2013.12.002

[28] Abreu MS, Giacomini AC, Gusso D, Rosa JG, Koakoski G, Kalichak F, Idalêncio R, Oliveira TA, Barcellos HH, Bonan CD, Barcellos LJ. Acute exposure to waterborne psychoactive drugs attract zebrafish. Comparative Biochemistry and Physiology. Toxicology & Pharmacology. 2016;**179**:37–43. DOI: 10.1016/j.cbpc.2015.08.009

[29] Howarth DL, Lindtner C, Vacaru AM, Sachidanandam R, Tsedensodnom O, Vasilkova T, Buettner C, Sadler KC. Activating transcription factor 6 is necessary and sufficient for alcoholic fatty liver disease in zebrafish. PLoS Genetics. 2014;**10**(5):e1004335. DOI: 10.1371/journal.pgen.1004335

[30] Nishino T, Takeuchi T, Takechi K, Kamei C. Evaluation of anxiolytic-like effects of some short-acting benzodiazepine hypnotics in mice. Journal of Pharmacological Sciences. 2008;**107**(3):349–354. DOI: 10.1254/jphs.08107FP

[31] Figueredo YN, Rodríguez EO, Reyes YV, Domínguez CC, Parra AL, Sánchez JR, Hernández RD, Verdecia MP, Pardo AGL. Characterization of the anxiolytic and sedative profile of JM-20: A novel benzodiazepine-dihydropyridine hybrid molecule. Neurological Research. 2013;**35**(8):804–812. DOI: 10.1179/1743132813Y.0000000216

[32] Giacomini AC, Abreu MS, Giacomini LV., Siebel AM, Zimerman FF, Rambo CL, Mocelin R, Bonan CD, Piato AL, Barcellos LJ. Fluoxetine and diazepam acutely modulate stress induced-behavior. Behavioural Brain Research. 2016;**296**:301–310. DOI: 10.1016/j.bbr.2015.09.027

[33] Egan RJ, Bergner CL, Hart PC, Cachat JM, Canavello PR, Elegante MF, Elkhayat SI, Bartels BK, Tien AK, Tien DH, Mohnot S, Beeson E, Glasgow E, Amri H, Zukowska Z, Kalueff AV. Understanding behavioral and physiological phenotypes of stress and anxiety in zebrafish. Behavioural Brain Research. 2009;**205**(1):38–44. DOI: 10.1016/j.bbr.2009.06.022G

[34] Maximino C, Silva AWB, Araujo J, Lima MG, Miranda V, Puty B, Benzecry R, Picanço-Diniz DLW, Gouveia A Jr, Oliverira KRM, Herculano. Fingerprinting of psychoactive drugs in zebrafish anxiety-like behaviors. PLoS One. 2014;**9**(7):e103943. DOI: 10.1371/journal.pone.0103943

[35] Gupta P, Khobragade SB, Shingatgeri VM, Rajaram SM. Assessment of locomotion behavior in adult Zebrafish after acute exposure to different pharmacological reference compounds. Drug Development and Therapeutics. 2014;**5**(2):127–33. DOI: 10.4103/2394-2002.139626

[36] Richendrfer H, Pelkowski SD, Colwill RM, Creton R. On the edge: Pharmacological evidence for anxiety-related behavior in zebrafish larvae. Behavioural Brain Research. 2012;**228**(1):99–106. DOI: 10.1016/j.bbr.2011.11.041

[37] Cerdà J, Conrad M, Markl J, Brand M, Herrmann H. Zebrafish vimentin: Molecular characterization, assembly properties and developmental expression. European Journal of Cell Biology. 1998;**77**(3):175-187. DOI: 10.1016/S0171-9335(98)80105-2

[38] Pinheiro-da-Silva J, Silva PF, Nogueira MB, Luchiari AC. Sleep deprivation effects on object discrimination task in zebrafish (*rerio*). Animal Cognition. 2017; **20**(2):159–169. DOI: 10.1007/s10071-016-1034-x

[39] Zuur AF, Ieno EN, Elphick CS. A protocol for data exploration to avoid common statistical problems. Methods in Ecology and Evolution. 2010;**1**(1):3–14. DOI: 10.1111/j.2041-210X.2009.00001.x

[40] Ripley B, Venables B, Hornik K, Gebhardt A, Firth D. Package 'MASS'[Internet]. 2013. Available from: http://www.trichord-inc.com/pricing/frames/content/solar_power.pdf [Accessed: 05-01-2017]

[41] Team RC. R: A language and environment for statistical computing [Internet]. R Foundation for Statistical Computing. 2015. Available from: https://www.r-project.org/ [Accessed: 05-01-2017]

[42] Lenth RV, Hervé M. lsmeans: Least-Squares Means. R package version 2.11 [Internet]. 2014. Available from: http//CRAN R-project org/package=lsmeans [Accessed: 05-01-2017]

[43] De Luca MA, Bassareo V, Bauer A, Di Chiara G. Caffeine and accumbens shell dopamine. Journal of Neurochemistry. 2007;**103**(1):157–163. DOI: 10.1111/j.1471-4159.2007.04754.x

[44] Richendrfer H, Pelkowski SD, Colwill RM, Créton R. Developmental sub-chronic exposure to chlorpyrifos reduces anxiety-related behavior in zebrafish larvae. Neurotoxicology and Teratology. 2012;**34**(4):458–465. DOI: 10.1016/j.ntt.2012.04.010

[45] Angelucci MEM, Cesário C, Hiroi RH, Rosalen PL, Da Cunha C. Effects of caffeine on learning and memory in rats tested in the Morris water maze. Brazilian Journal of Medical and Biological Research. 2002;**35**(10):1201–1208. DOI: 10.1590/S0100-879X2002001000013

[46] Dunwiddie TV, Masino SA. The role and regulation of adenosine in the central nervous system. Annual Review of Neuroscience. 2001;**24**(1):31–55. DOI: 10.1146/annurev.neuro.24.1.31

[47] Daly JW, Fredholm BB. Caffeine–an atypical drug of dependence. Drug and Alcohol Dependence. 1998;**51**(1–2):199–206. DOI: 10.1016/S0376-8716(98)00077-5

[48] Kirch DG, Taylor TB, Gerhardt GA, Benowitz NL, Stephen C, Wyatt RJ. Effect of chronic caffeine administration oh monoamine and monoamine metabolite concentratiows in rat brain. Neuropharmacology. 1990;**29**(6):599–602. DOI: 10.1016/0028-3908(90)90073-Z

[49] Nehlig A. Is caffeine a cognitive enhancer. Journal of Alzheimer's Disease. 2010;**20**(S1): S85–94. DOI: 10.3233/JAD-2010-091315.

[50] Ribeiro JA, Sebastiao AM. Caffeine and adenosine. Journal of Alzheimer's Disease. 2010;**20**(S1):3–15. DOI: 10.3233/JAD-2010-1379.

[51] Abreu RV, Silva-Oliveira EM, Moraes MF, Pereira GS, Moraes-Santos T. Chronic coffee and caffeine ingestion effects on the cognitive function and antioxidant system of rat brains. Pharmacology Biochemistry and Behavior. 2011;**99**(4):659–664. DOI: 10.1016/j.pbb.2011.06.010

[52] de Aguiar RA, Turnes T, Cardoso TE, Vasconcelos DIC, Caputo F. Efeito da ingestão de cafeína em diferentes tarefas de tempo de reação. Revista Brasileira de Ciências do Esporte. 2011;**34**(2):465–476. DOI: 10.1590/S0101-32892012000200015.

[53] Santos LC, Ruiz-oliveira J, Oliveira JJ, Silva PF, Luchiari AC. Pharmacology, biochemistry and Behavior. Irish coffee: Effects of alcohol and caffeine on object discrimination in zebrafish. Pharmacology Biochemistry and Behavior. 2016;**143**:34–43. DOI: 10.1016/j.pbb.2016.01.013

[54] Glade MJ. Caffeine–not just a stimulant. Nutrition. 2010;**26**(10):932–938. DOI: 10.1016/j.nut.2010.08.004

[55] Burgess JW. Chronic exposure to caffeine during early development modifies spatial behavior in juvenile jewel fish schools. Pharmacology Biochemistry and Behavior. 1982;**17**(1):137–140. DOI: 10.1016/0091-3057(82)90275-1

[56] Nehlig A, Daval JL, Debry G. Caffeine and the central nervous system: Mechanisms of action, biochemical, metabolic and psychostimulant effects. Brain Research Reviews. 1992;**17**(2):139–170. DOI: 10.1016/0165-0173(92)90012-B

[57] Ardais AP, Borges MF, Rocha AS, Sallaberry C, Cunha RA., Porciúncula LO. Caffeine triggers behavioral and neurochemical alterations in adolescent rats. Neuroscience. 2014;**270**:27–39. DOI: 10.1016/j.neuroscience.2014.04.003

[58] Hughes RN, Hancock NJ. Strain-dependent effects of acute caffeine on anxiety-related behavior in PVG/c, Long–Evans and Wistar rats. Pharmacology Biochemistry and Behavior. 2016;**140**:51–61. DOI: 10.1016/j.pbb.2015.11.005

[59] Jabbar SB, Hanly MG. Fatal caffeine overdose: A case report and review of literature. American Journal of Forensic Medicine and Pathology. 2013;**34**(4):321–324. DOI: 10.1097/PAF.0000000000000058

[60] Rodriguez RS, Haugen R, Rueber A, Huang CC. Reversible neuronal and muscular toxicity of caffeine in developing vertebrates. Comparative Biochemistry and Physiology Part C: Toxicology & Pharmacology. 2014;**163**:47–54.

Chapter 5

Development of Tumor-Specific Caffeine-Potentiated Chemotherapy Using Span 80 Nano-Vesicles DDS

Tatsuhiko Miyazaki, Hiroshi Nakata and Keiichi Kato

Additional information is available at the end of the chapter

http://dx.doi.org/10.5772/intechopen.69155

Abstract

Osteosarcoma cases with metastasis have poor prognosis in general. Recently, caffeine-potentiated chemotherapy, which is chemotherapy with caffeine dosage against malignancies, has manifested potently high efficacy as well as diverse effects. Recently, we demonstrated that nonionic vesicles prepared from Span 80 have promising physicochemical properties, which let them an attractive option besides the common liposomes. Here, we manifested the tumor-specific caffeine-potentiated chemotherapy against osteosarcoma in murine model employing a novel drug delivery system (DDS) with Span 80 nano-vesicles. C3H/HeJ mice underwent transplantation of LM8 osteosarcoma cell line and then were doped with therapeutic agents. Caffeine was employed as an enhancer in addition to ifosfamide (IFO) as the antitumor agent. in vitro, the united administration of IV + CV revealed significant induction of tumor apoptosis in the early phase. In vivo study manifested that IV + CV-administration markedly decreased the tumor volume as well as the viable tumor area than in the other groups. No marked organ damage was observed in the IV or IV + CV groups as well as fertility injury and/or malformations in their progeny. This novel DDS might have the importance for clinical application in primary tumors as well as the metastatic osteosarcoma.

Keywords: DDS, Span 80 nano-vesicles, caffeine-potentiated chemotherapy, mouse model

1. Introduction

At present, osteosarcoma cases with metastasis, especially in lungs, have poor prognosis [1, 2]. In recent years, caffeine-potentiated chemotherapy, which is chemotherapy with caffeine dosage against malignancies, has manifested potently high efficacy [3, 4]. Nevertheless, this method may induce adverse effects that patients suffer, which include most commonly tachycardia, nausea, psychiatric symptoms, as well as lethal arrhythmia with individual diversity [5]. On the other hand, there have been numerous developments in novel drug delivery systems (DDSs) for drug carriers for the treatment of various diseases that enabled target-specific drug delivery resulting in the prevention of side effects [6–8].

Recently, we demonstrated that nonionic vesicles prepared from Span 80 have promising physicochemical properties, such as high membrane fluidity associated with low-phase transition temperature, which make them an attractive possible alternative to the commonly used liposomes. Lipid vesicles have been extensively studied. Since the discovery of mechanism for liposome by Bangham et al. that aqueous phase of phosphatidylcholines includes self-closed phospholipid bilayers, which can capture and obtain water-soluble molecules [9], lipid vesicles have been actively investigated. Following early reports on vesicle formation from completely synthetic amphiphiles [10], vesicles have been prepared from a large number of different surfactants [11, 12]. Many vesicle systems have been characterized to some extent and applied in various research areas, ranging from pharmaceuticals [12–15], food technology [12, 16, 17], and analytical applications to origin-of-life [18, 19] and artificial cell studies [20]. Vesicles based on nonionic surfactants (so-called "niosomes") [12, 21] were first used in the cosmetic industry [10, 21, 22] as alternatives to phospholipid-based vesicles (liposomes). One of the many surfactants used for niosome preparations is Span 80, a cheap, molecularly heterogeneous nonionic surfactant that is also applied as food emulsifier and in oral pharmaceuticals [23, 24]. Span 80 is known as sorbitan mono-oleate generally; nevertheless, commercially available Span 80 may be a heterogeneous mixture of sorbitan mono-, di-, tri, and tetra-esters which could let high fluidity and vascular permeability [25–27].

A successful therapeutic murine model of transplanted colon cancer employs the DDS using Span 80 vesicles which have immobilized polysaccharides [28]. In this chapter, we introduce a novel DDS by using Span 80 nano-vesicles, and manifested that tumor-specific caffeine-potentiated chemotherapy for murine osteosarcoma using a novel DDS with Span 80 nano-vesicles showed significant antitumor effects, as well as limited adverse effects.

2. Span 80 nano-vesicles

2.1. Characteristics of Span 80 as a material for food and pharmaceuticals

As mentioned above, Span 80 is known as sorbitan mono-oleate; nevertheless, commercially available Span 80 might be heterogeneous mixture, rather mainly diesters, in addition to triesters and tetraesters [23, 25]. Furthermore, the polar headgroup of the different esters present in Span 80 is not sorbitol, but more likely one of the different forms of anhydrized sorbitol [29, 30], a cheap, molecularly heterogeneous nonionic surfactant that is also applied as food emulsifier and in oral pharmaceuticals [31]. The substantial molecular heterogeneity of commercial Span 80 is (i) a consequence of the conditions used for the synthesis (reaction of sorbitol with fatty acids (mainly oleic acid) at elevated temperature) [24, 29] and (ii) based on the fact that there are no purification method following the synthesis with excellent cost-performance ratio for yielding inexpensive products and applicable to large-scale purification. Commercially available Span 80 was determined by its molecular composition. The property of it may even be better when compared to the properties of the individual purified components of commercially available Span 80.

2.2. Preparation of Span 80 nano-vesicles in different forms

In our study, we prepared several variations of Span 80 vesicles as follows: Commercially available Span 80 was processed using the two-step emulsification method (Span 80 vesicles type 1), sequentially extruded by a polycarbonate membrane (Span 80 vesicles type 2) or ultrafiltrated (Span 80 vesicles type 3). Fractionation of commercially available Span 80 by chromatography into the different ester groups (see **Figure 1**) and vesicular preparation using the defined mixture of the four kinds of different ester groups (Span 80 vesicles type 4) or preparation from the diester fraction (Span 80 vesicles type 5) were performed [25].

2.3. Evaluation of diameter and homology by a dynamic light scattering

Vesicle characterization by dynamic light scattering (DLS) and electron microscopy was performed. The different types of Span 80 vesicles prepared either in distilled water or in phosphate-buffered saline (PBS) solution were analyzed after vesicle preparation by dynamic light scattering. As expected from the different preparation methods used, the average vesicle size and size distribution depend on the vesicle type, independent of whether PBS or distilled water was used as aqueous medium. (a) The vesicle size would depend on the employed method; therefore, the prepared Span 80 vesicles should be kinetically trapped aggregates and might not have thermodynamically equilibrium structures, just like most of phosphatidylcholine vesicles. (b) The most homogeneous vesicles with the lowest polydispersity index were prepared by the extrusion method. When employing PBS as the aqueous medium, the size of Span 80 vesicle type 1 with an apparent size of 250 ± 45 nm, which was obtained by DLS, could be reduced into 105 ± 13 nm by extrusion through polycarbonate membranes with a nominal pore diameter of 100 nm. Extrusion method enabled to make more homogeneous vesicles with less polydispersity index. Span 80 vesicles type 2 manifested the appropriate diameter (c.a. 100 nm) for the drug delivery; therefore, these types of vesicles were employed for further analyses in the development of DDS.

2.4. Evaluation of diameter and physicochemical property by electron microscopy

Span 80 vesicles type 2 (100 nm) prepared in PBS was also statistically analyzed by cryo-transmission electron microscopy (cryo-TEM, **Figure 1A**), yielding a number-weighted average vesicle size of 63. This value is lower than the *z*-average value (scattering intensity weighted) determined by DLS. Next, the hydrodynamic diameter was evaluated; on the other hand, in cryo-TEM images, the projected, "true" size of a spherical vesicle was obtained. The discrepancy among these methods could be addressed that the electron dense headgroup area made vesicles more boundary. Conclusively, electron microscopy revealed the diameter and bilayer thickness of the vesicles by Cryo-TEM, as well as the presence of uniformity of the vesicles by freeze fracture

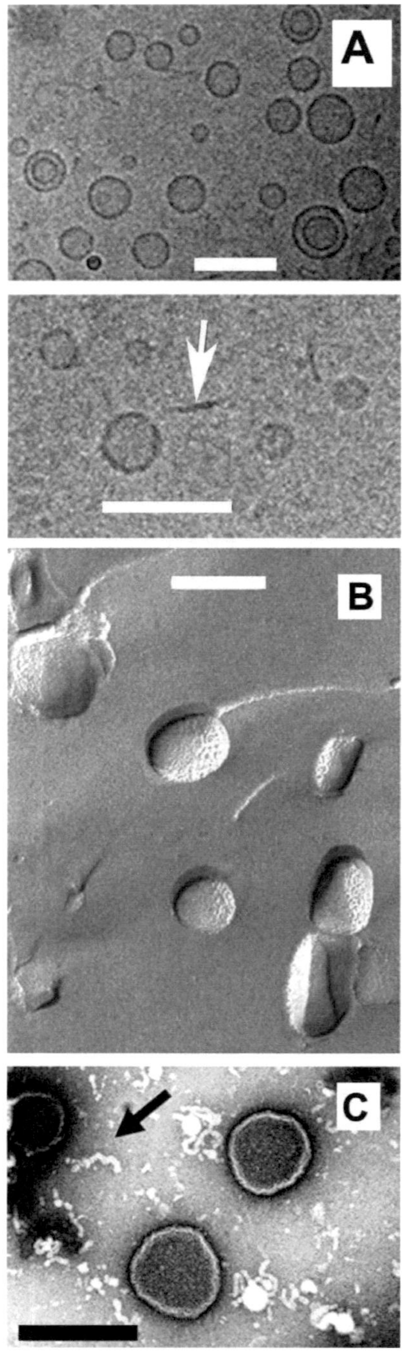

Figure 1. Electron microscopic analysis of Span 80 vesicles type 2 (100 nm), prepared in PBS. (A) Cryo-TEM micrographs showing unilamellar- and bilamellar vesicles and bilayer fragments (arrow). Length of the bar: 100 nm. Freeze fracture (B) and negative-staining (C) electron micrographs of Span 80 vesicles type 2 (100 nm), prepared in PBS solution. Length of the bar: 100 nm. The arrow in C points to one of the fragments present. [Reprinted with permission from Ref. [25]. Copyright (2008) American Chemical Society].

scanning electron microscopy (**Figure 1**). Electron micrographs revealed that Span 80 vesicle suspensions contain not only vesicles but also bilayer fragments. The clear contrast was observed among 1-palmitoyl-2-oleoyl-sn-glycero-3-phosphocholine (POPC) or dioleoylphosphatidylcholine (DOPC) vesicles which might be due to the molecular heterogeneity of Span 80 vesicles.

2.5. Temperature sensitivity of Span 80 vesicles

The temperature sensitivity of Span 80 vesicles might not link directly to the T_m value, because the observed fusion phenomenon did not develop at T_m as the temperature-sensitive vesicles based on 1,2-dipalmitoyl-sn-glycero-3-phosphocholine (DPPC) [32]. In DPPC-based vesicles, the thermos-responsive effect is the leakage of the aqueous contents when the temperature reaches T_m [33].

The Span 80 vesicles develop fusion in response to an increase in temperature. Therefore, the molecular mechanisms of thermos-response in each kind of vesicles are different. The presence of bilayer fragments in fused Span 80 vesicles at an elevated temperature is not clear at this moment. However, the nonionic headgroup in Span 80 could be dehydrated and result in vesicle-vesicle aggregation and fusion at the temperature above T_m.

2.6. Vesicle membrane fusion property

The vesicle fusion property may be advantageous for efficient drug delivery, and applications of the several types of Span 80 vesicles described and characterized in this chapter, and previous papers largely depend on the vesicle's cytotoxicity. Although previous studies of Span 80-based vesicles regarding cytotoxicity either as a drug carrier or as a gene vector were successful, further studies have been required before any conclusions with respect to pharmaceutical applications [25] can be drawn. Among the various types of Span 80 vesicles investigated, Span 80 vesicles type 2 (100 nm) might be the most attractive one (straightforward methodology with the requirement of simple equipment only).

3. Caffeine-potentiated chemotherapy using Span 80 nano-vesicles' DDS

We developed the murine osteosarcoma therapeutic model of caffeine-potentiated chemotherapy. In this model, as the therapeutic agents, ifosfamide (IFO) was employed as well as caffeine sodium benzoate (CSB) as an enhancer. As the murine osteosarcoma therapeutic model, C3H/HeJ mice underwent transplanted murine osteosarcoma cell line LM8. The detailed procedures were described in the original paper [34].

3.1. Preparation of Span 80 nano-vesicles

Span 80 nano-vesicles, which contained IFO and/or caffeine, were freshly prepared as previously described [28]. Briefly, materials for assembling nano-vesicles containing Span 80 and Tween-80 [35], cholesterol, which worked as the stabilizer of the membrane, polyethylene glycol, used as a stealth modifier against macrophages [28], and the solvents, normal hexane and normal saline, were purchased, respectively. All processes were performed under sterilized conditions.

The two-step emulsification method was employed to process and purify the nano-vesicles. Span 80 and cholesterol were dissolved in hexane by homogenization with a micro-homogenizer in a sterilized brown glass bottle. Sequentially, the first emulsion was prepared by adding IFO and/or CSB, which dissolved in normal saline into the Span 80 material followed by homogenization. As a negative control, phosphate-buffered saline was dripped alternatively. The second-stage emulsion was processed by evaporation using a rotary vacuum evaporator on a water bath at 37°C followed by homogenization with Tween-80.

The second emulsion was centrifuged using ultra-centrifugation equipment. After aspiration of the supernatant, the sediment of the Span 80 vesicles was weighed and then suspended in normal saline at a concentration of 20% w/v. By this method, IFO Span 80 vesicles (IV), CSB Span 80 vesicles (CV), and PBS-alone Span 80 vesicles (PV) were prepared. Immediately before the use *in vivo* or *in vitro*, these suspensions became extruded by a custom-made extruder with a drain disk of 100-μm thickness and a Nucleopore membrane® of 100-nm pore size to control the vesicular size. As a result, the diameter of the vesicles was evaluated by the dynamic light-scattering device and revealed 117 nm at average.

3.2. In vitro evaluation of the antitumor effects of the nano-vesicles

The murine osteosarcoma cell line, LM8, was obtained and employed as an osteosarcoma model. LM8 cells in Dulbecco's modified Eagle's medium were plated and cultured in 24-well culture dishes for a few days until the cells showed semi-confluent state. Next, antitumor agents with or without Span 80 vesicles, including PV, IV, CV, direct administration of IFO or CSB, as well as the combination of IV + CSB, IFO + CV and IV + CV were administered. Cells were incubated with the antitumor agents at 37°C for 1 or 2 h, and then the cells were harvested and evaluated for apoptosis and cell viability, respectively.

In vitro analyses revealed cultured LM8 murine osteosarcoma cells with IV + CV almost complete cell death by the trypan blue assay, on the other hand, PBS, CSB, PV, and CV manifested almost no cell death. IFO resulted in 13%, IV resulted in 28%, and IFO + CV resulted in 75% cell death (**Figure 2** and **Table 1**).

3.3. Apoptosis detection by propidium iodide (PI) method

Briefly, cells were suspended in ice-cold Hank's balanced saline solution (HBSS), followed by fixation with 70% EtOH at −20°C [36]. Fixed cells were centrifuged, then pellets were re-suspended in extraction buffer of pH 7.8 which contained Na_2HPO_4, citric acid, and 0.1% Triton X-100 at 37°C. Then, a staining solution of pH 6.8 containing PIPES, NaCl, Mg_2Cl, Triton X-100, PI, and 50 RNAse H was added to the cell suspension, and the fluorescence intensity was evaluated and analyzed in triplicate by the FACStation® and CellQuest® software.

PI analyses revealed that almost cell population (97%) underwent apoptotic cell death, which treated with IV + CV. By contrast, PBS, CSB, PV, and CV conducted cell death in very limited population, while IFO and IV let the small population into apoptosis and/or necrosis (8.8–10.2%), as well as IFO + CSB and IV + CSB induced increasing cell death to approximately a quarter to one-third of the population (**Figure 3** and **Table 2**).

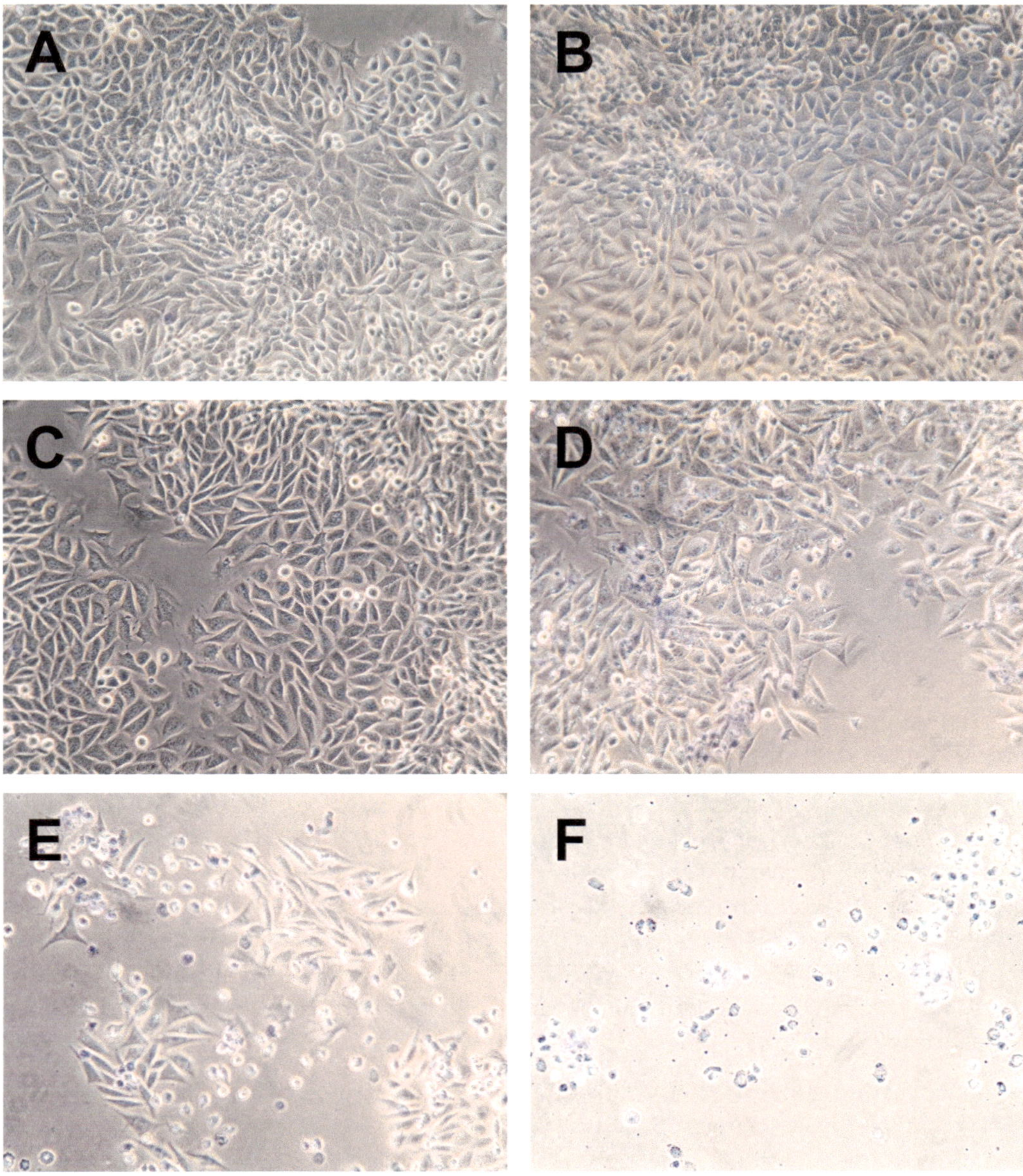

Figure 2. Representative photomicrographs of the trypan blue-stained LM8 cells after a 2-h incubation with antitumor agents: (A) PV, (B) CV, (C) IFO, (D) IV, (E) IFO + CV, and (F) IV + CV. [Reprinted with permission from Ref. [34]. Copyright (2014) Spandidos Publications].

3.4. Murine osteosarcoma therapeutic model

For the therapeutic model, C3H/HeJ mice were employed because they are H2-matched to LM8 osteosarcoma cell line since this cell line was originated from that strain of mouse [37]. LM8 cells (3.0×10^6 cells per mice) were subcutaneously transplanted into 6-week-old male

Treatment	Population of nonviable cells (%) (mean ± SD)
PBS	1.5 ± 0.9
CSB	2.1 ± 1.2
PV	3.3 ± 1.8
CV	3.1 ± 1.9
IFO	13 ± 3.4[a]
IV	28 ± 5.5[a]
IFO + CSB	25 ± 6.7[b]
IV + CSB	40 ± 9.2[c]
IFO + CV	75 ± 10.8[d]
IV + CV	98 ± 1.2[e]

[a]$P < 0.05$ versus PBS, CBS, PV, and CV.
[b]$P < 0.05$ versus PBS, CSB, PV, CV, and IFO.
[c]$P < 0.01$ versus IFO, $P < 0.05$ versus IV and IFO + CSB.
[d]$P < 0.01$ versus IV + CSB and the other groups
[e]$P < 0.05$ versus IFO + CV, $P < 0.01$ versus the other groups.

Table 1. Nonviable cell population in trypan blue analysis.

C3H/HeJ mice. After c.a. 3 weeks, when the tumor volume reached up to 500 mm^3, injection of the therapeutic agents was started. The administration protocol is schemed in **Figure 4**.

The agents were administered individually or in combinations as follows: PBS (i.v., sham administration), CV (0.1 mg/g BW), IFO (direct i.v. 0.1 mg/g BW), IV (i.v. 0.1 mg/g BW), IV + CSB, and IV + CV. Five to eight animals in each groups were analyzed in the study. The therapeutic agents were intravenously administered via tail vein on days 0, 2, and 4, followed by the harvest under anesthesia on day 7. Tumor diameter as well as the body weight of each individual was measured every day. At the time of the harvest, volumes and weights of tumors were evaluated; subsequently, the entire organs and the tumors were processed for histopathological analyses.

No significant differences were noted in body weights among each other in the groups. It was marked that the tumor volumes in the IV + CV group were reduced as compared to those of the control groups (PBS and CV), as well as a tendency toward a decrease against the PV- and IFO-direct i.v. groups on days 5–7 (**Figure 5**) could be shown.

3.5. Histopathological analyses

The histopathological analyses of the harvested tumors and entire organs were executed on the formalin-fixed, paraffin-embedded tissue section. The area of viable tumor was evaluated as the viability of the tumor tissue in hematoxylin-eosin (HE)-stained sections. Next, in order to determine the adverse effects, entire organ tissue sections stained with HE, periodic acid-Schiff (PAS), and Elastica-Goldner stains were accessed by skilled pathologists.

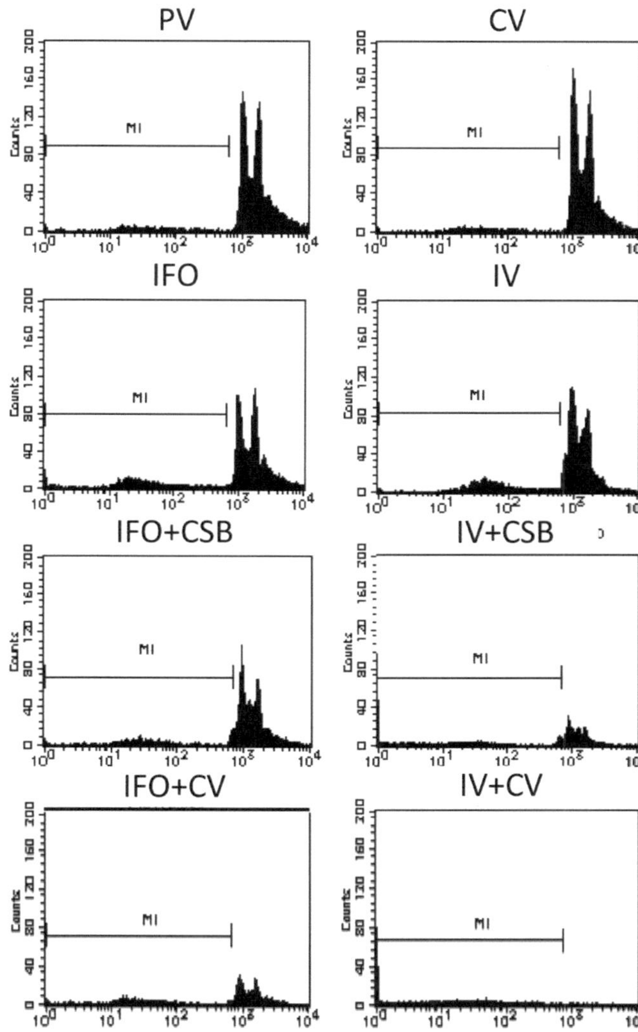

Figure 3. The PI-staining apoptosis assay using flow cytometry. Each panel shows the event count (vertical axis) at each intensity (horizontal axis). The population of apoptotic and/or necrotic cells was measured as M1 and was shown in Table 2. [Reprinted with permission from Ref. [34]. Copyright (2014) Spandidos Publications].

IFO, IV, IV + CSB, and IV + CV groups revealed significantly smaller viable tumor areas in comparison to the controls. Moreover, the IV + CV group revealed markedly reduced viable tumor areas against the IFO and IV groups (**Figures 6** and **7**).

3.6. Histopathological analyses for adverse effects in vivo

To determine whether the DDS using Span 80 vesicles could prevent or reduce hazardous adverse effects due to the chemotherapeutic agents, the entire organs were histopathologically examined. Marked prevention of adverse effect was histopathologically

Treatment	Population of nonviable cells (%) (mean ± SD)
PV	1.6 ± 1.1
CV	1.4 ± 1.0
IFO	8.8 ± 1.9[a]
IV	10.2 ± 2.9[a]
IFO + CSB	16.5 ± 3.9[b]
IV + CSB	25.2 ± 4.2[b]
IFO + CV	32.8 ± 5.9[b]
IV + CV	97.1 ± 1.9[c]

[a] $P < 0.05$ versus PV and CV.
[b] $P < 0.05$ versus PV and CV.
[c] $P < 0.001$ versus PV and CV.

Table 2. Population of nonviable cells (M1) in flow cytometry.

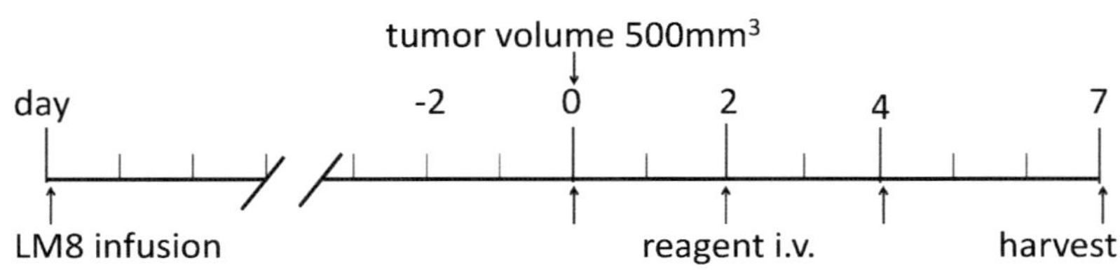

Figure 4. The in vivo therapeutic model. Administration of the antitumor agents was initiated when the tumor volume reached ~500 mm³ (day 0), and continued on days 2 and 4. Then the mice were sacrificed and analyzed on day 7. [Reprinted with permission from Ref. [34]. Copyright (2014) Spandidos Publications].

observed in the kidney, liver, and testis. Significant tubular injury, which was recognized as a loss of brush border, as well as the glomerular damages such as the expansion of the mesangial matrix was manifested in the IFO-direct i.v. group in contrast to those in the IV and/or IV+CV groups (**Figure 8A** and **B**). Furthermore, in the liver, spotty or grouping necrosis as well as reduced glycogen storage in the hepatocytes was observed in the IFO-direct i.v. groups in contrast to those in the IV and IV + CV groups (**Figure 8C** and **D**). Moreover, the IV and IV + CV groups manifested no remarkable changes in spermatogenesis, while the IFO-direct i.v. group revealed marked suppression of spermatogenesis along with the necrosis of the germ cells (**Figure 8E** and **F**).

3.7. Fertility test

In order to elucidate whether the DDS with Span 80 nano-vesicles could be able to prevent the infertility, fertility tests were performed. Three male C3H/HeJ mice in each group that were administered IFO, IV, or IV + CV were cross-mated with 6-week-old female C3H/HeJ mice individually; then the fertility of each male was evaluated.

http://dx.doi.org/10.5772/intechopen.69155

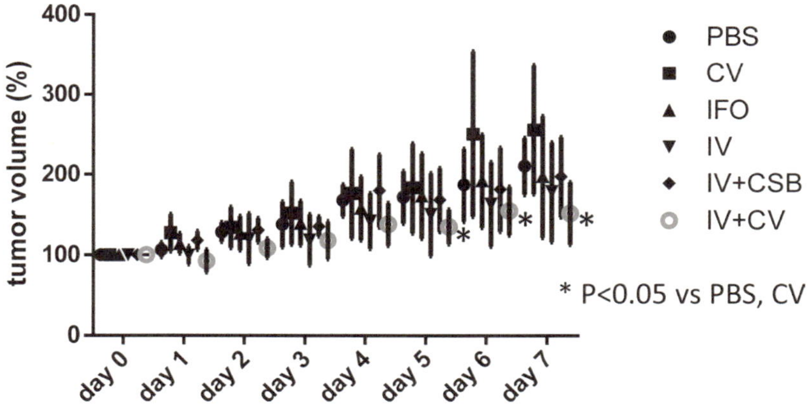

Figure 5. Trace of tumor volumes after antitumor agent administration. The symbols represent the mean value of each group, and the bars represent standard deviation. [Reprinted with permission from Ref. [34]. Copyright (2014) Spandidos Publications].

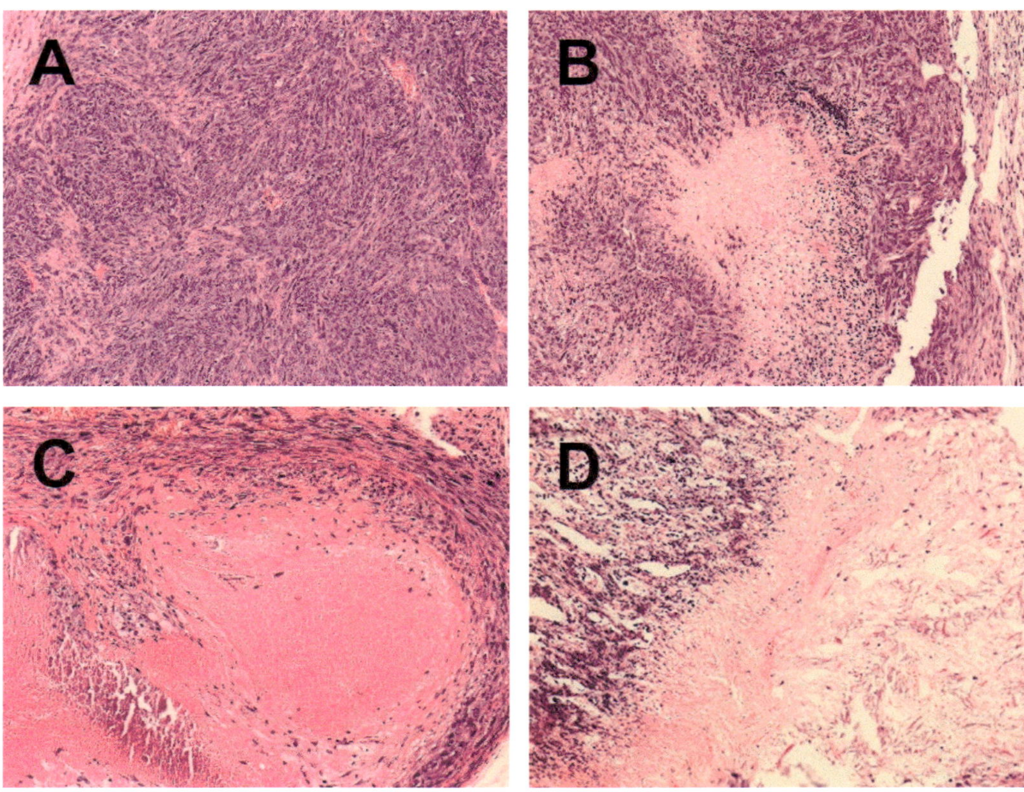

Figure 6. Representative photomicrographs of the tumors treated with (A) CV, (B) IFO, (C) IV, and (D) IV + CV. [Reprinted with permission from Ref. [34]. Copyright (2014) Spandidos Publications].

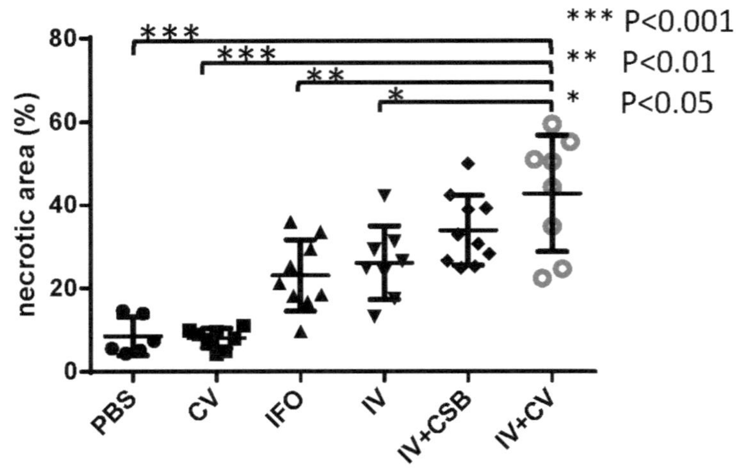

Figure 7. The nonviable tumor area (%) in each animal from each group. The center bars express mean value, as well as the upper and lower bars express standard deviation. [Reprinted with permission from Ref. [34]. Copyright (2014) Spandidos Publications].

The test revealed that the male mice after IV + CV administration had normal fertility, and there were no malformations in their progeny.

4. Discussion and conclusion

A promising suggestion from the therapeutic model of the DDS with Span 80 vesicles was conducted that this DDS could enhance the therapeutic effects of IFO and caffeine-potentiated IFO chemotherapy, over and above prevent the hazardous adverse effects induced by chemotherapy. In vitro studies revealed drastic cell death in a very early phase by IV + CV administration in contrast to the "mild" apoptotic cell death inference by the administration of IFO alone, IV alone, or combinations of IFO + CSB, IV + CSB, and IFO + CV. These findings suggested that the immediate delivery of therapeutic agents into the cytosol by IV + CV addition might induce extremely rapid apoptosis. Fusion of Span 80 vesicles and cell membrane could be implicated in this rapid response; nevertheless, the comprehensive mechanisms are still unknown.

Marked development of the DDS employing nano-vesicles has been reported along with the development of many types of phospholipids and/or detergents [6–8, 38]. The Span 80 nanovesicle might be a promising material among them, based on its favorable physicochemical properties, including membrane fluidity and flexibility. With respect to membrane fluidity, Hayashi et.al. reported that Span 80 vesicles have markedly high fluidity with various cholesterol contents in comparison to conventional phospholipid liposomes, such as 1,2-dipalmitoyl-sn-glycero-3-phosphocholine and 1-palmitoyl-2-oleoyl-sn-glycero-3-phosphocholine liposomes. Not only the high fluidity, also Span 80 vesicles manifest much more flexibility in comparison to DPPC and POPC liposomes [26].

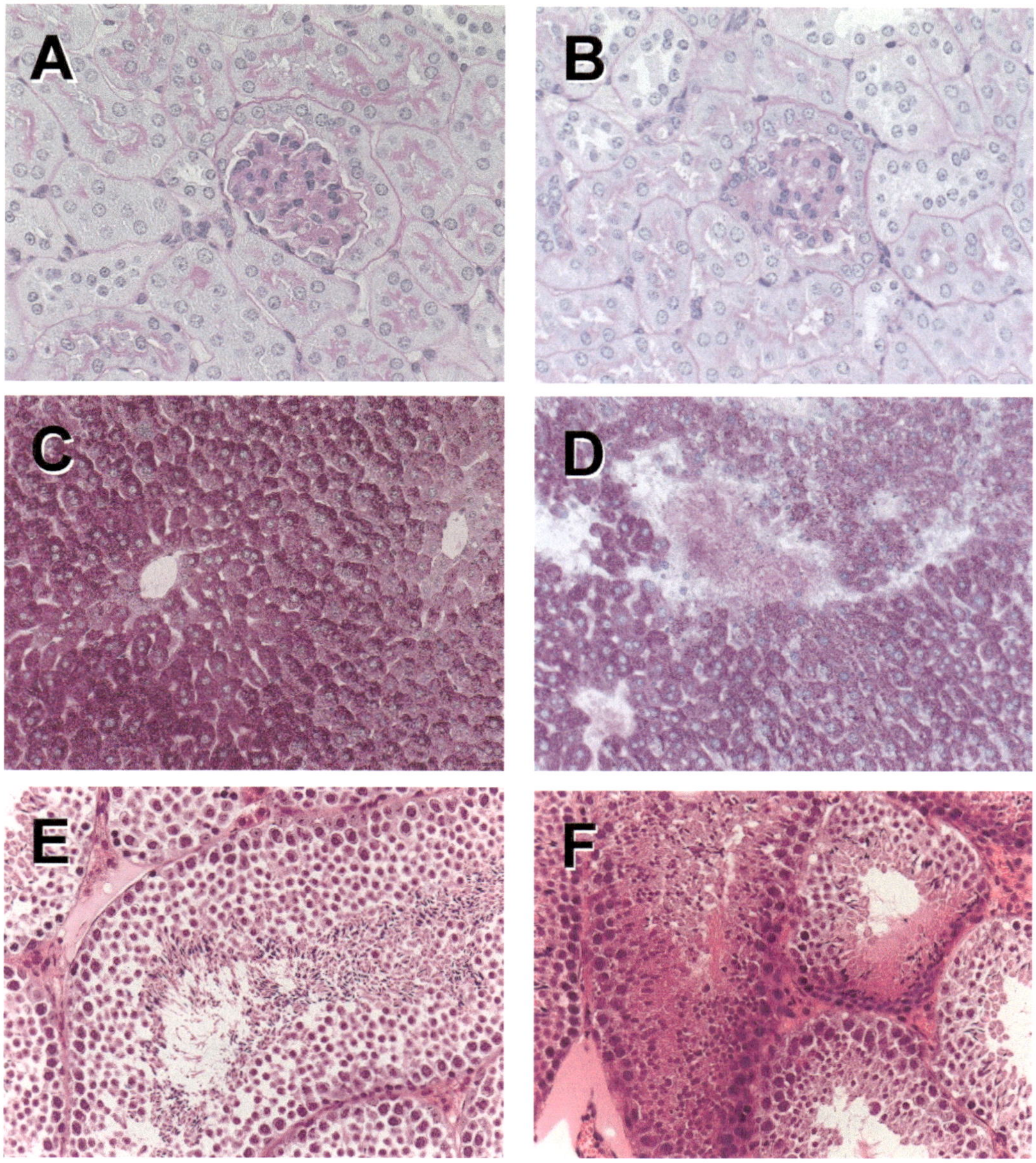

Figure 8. Representative histopathological characteristics of the renal tissue (A and B), liver (C and D), and testis (E and F), which were harvested from the animals with IV+CV (A, C, and E) and IFO i.v. (B, D, and F) administration.

Nonvesicular aggregates are observed often in the common liposome suspensions; on the other hand, Span 80 vesicle suspensions also could contain limited amount of nonvesicular aggregates such as tubulin structures. Recently, Kato et al. manifested that the Span 80 nano-vesicle might be a kind of kinetically trapped aggregates and might not have thermodynamic equilibrium structures, like in most kinds of vesicles prepared from phosphatidylcholines (liposomes) [25].

The adverse effects induced by ifosfamide have been reported in kidney [35, 39–42], liver [43–45], gonadal cells [46–48], and bone marrow more frequently to other organs [49, 50]. In the therapeutic model of Span 80 DDS, the mice in the IFO-direct i.v. group also manifested moderate tubular injury as well as the glomerular damage in kidney, moreover, severe inhibition of spermatogenesis with gonadal cell necrosis. On the other hand, the novel DDS employing Span 80 nano-vesicles manifested marked prevention of the hazardous adverse effects in the kidney, liver, and testis. These favorable results could implicate the tumor selectivity of the Span80 vesicles, which might be at least partially resulting from the refraining from phagocytosis taken on the pegylation of the vesicles and also possibly on the lower permeability at the blood-testis barrier in comparison to the direct injection of low-molecular-weight molecules such as IFO [48, 51].

The results of our study indicated that higher vascular permeability and inclination to fuse with the instable cell membrane of the tumors based on high fluidity and flexibility as well as pegylation could result in the higher tumor selectivity of Span 80 vesicles [52]. Recently, a cell fusion model using Span 80 vesicles has been reported [27]. Furthermore, our results manifested that CV conducted markedly better enhancement of antitumor effects than that of the direct injection of CSB. This might be addressed by the pegylation-associated tumor selectivity as well as the inclination for cell fusion which might enable the immediate caffeine delivery into the cytoplasm. Moreover, the prevention of caffeine toxicity, which causes the withdrawal of numerous patients from caffeine-potentiated chemotherapy, could be prevented based on the selectivity of caffeine delivery by using Span 80 vesicles [5]. Currently, a DDS of doxorubicin containing liposomal nano-vesicles is applied in actual cancer therapy with marked efficacy [6–8, 53–55]. The next-generation liposomes with membrane-bound-targeting molecules have also been under development. An anticarcinoma application of Span 80 vesicles containing doxorubicin with or without membrane-bound-targeting molecule was recently reported [52]. As described above, Span 80 has favorable physicochemical properties; moreover, it also confirmed risk-free information because it has been already used as a stabilizer for injected drugs. Furthermore, the cost of Span 80 vesicles should be drastically cheaper than common liposomes. Therefore, the DDS with Span 80 nano-vesicles might be a promising next-generation DDS.

Recently, a novel treatment method for lymph node metastasis using a lymphatic drug delivery system with nano-/microbubbles has been advocated [56–58]. Those reports suggested that the lymphatic DDS might drastically improve the tissue selectivity and response rates to the metastatic tumors which had been limited in the hematogenous administration of drugs resulting in poor prognosis. Furthermore, those models could prevent the systemic toxic effects of the treatment; nevertheless, they employed highly toxic doxorubicin as the antitumor agent. The caffeine-potentiated chemotherapy employing the DDS with Span 80 vesicles might have excellent affinity to this lymphatic administration and more effective and less harmful treatment onto the tumor with lymph-nodal metastasis.

In conclusion, the DDS with Span 80 vesicles may enhance the antitumor effects of IFO and of caffeine-potentiated IFO chemotherapy against osteosarcoma. Moreover, the usage of this DDS may suppress the adverse effects, which were induced by the chemotherapy. Thus, this

DDS model has promising importance for clinical application in the therapy of metastatic osteosarcoma as well as the primary tumors.

Acknowledgements

We would like to thank Dr Keita Hayashi (Nara National College of Technology) and Dr Yosuke Omokawa (Nanocareer Inc. Ltd.) for their kind technical support, permission of documentation of their data, and useful advice. We are also deeply thankful to Professor Peter Walde (ETH, Zürich, Switzerland) for his helpful scientific discussion and advice. This study was supported by JSPS Kakenhi Grant Numbers 23592191, 25460496, and 16K08735.

Author details

Tatsuhiko Miyazaki[1]*, Hiroshi Nakata[2] and Keiichi Kato[3]

*Address all correspondence to: tats_m@gifu-u.ac.jp

1 Pathology Division, Gifu University Hospital, Gifu, Japan

2 Department of Orthopaedic Surgery, HITO Hospital, Shikokuchuo City, Japan

3 Ehime University Graduate School of Science and Engineering, Matsuyama, Japan

References

[1] Wu PK, Chen WM, Chen CF, Lee OK, Haung CK, Chen TH. Primary osteogenic sarcoma with pulmonary metastasis: Clinical results and prognostic factors in 91 patients. Japanese Journal of Clinical Oncology. 2009;**39**(8):514–522. DOI: 10.1093/jjco/hyp057

[2] Clark JC, Dass CR, Choong PF. A review of clinical and molecular prognostic factors in osteosarcoma. Journal of Cancer Research and Clinical Oncology. 2008;**134**(3):281–297. DOI: 10.1007/s00432-007-0330-x

[3] Kimura H, Tsuchiya H, Shirai T, Nishida H, Hayashi K, Takeuchi A, Ohnari I, Tomita K. Caffeine-potentiated chemotherapy for metastatic osteosarcoma. Journal of Orthopaedic Science. 2009;**14**(5):556–565. DOI: 10.1007/s00776-009-1372-5

[4] Tsuchiya H, Tomita K, Mori Y, Asada N, Morinaga T, Kitano S, Yamamoto N. Caffeine-assisted chemotherapy and minimized tumor excision for nonmetastatic osteosarcoma. Anticancer Research. 1998;**18**(1B):657–666

[5] Hayashi K, Tsuchiya H, Yamamoto N, Shirai T, Yamauchi K, Takeuchi A, Kawahara M, Miyamoto K, Tomita K. Impact of serum caffeine monitoring on adverse effects and

chemotherapeutic responses to caffeine-potentiated chemotherapy for osteosarcoma. Journal of Orthopaedic Science. 2009;**14**(3):253–258. DOI: 10.1007/s00776-009-1336-9

[6] Yang C, Fu ZX. Liposomal delivery and polyethylene glycol-liposomal oxaliplatin for the treatment of colorectal cancer (review). Biomedical Reports. 2014;**2**(3):335–339. DOI: 10.3892/br.2014.249

[7] Brochu H, Polidori A, Pucci B, Vermette P. Drug delivery systems using immobilized intact liposomes: A comparative and critical review. Current Drug Delivery. 2004;**1**(3): 299–312

[8] Sultana S, Khan MR, Kumar M, Kumar S, Ali M. Nanoparticles-mediated drug delivery approaches for cancer targeting: A review. Journal of Drug Targeting. 2013;**21**(2):107–125. DOI: 10.3109/1061186X.2012.712130

[9] Bangham AD, Standish MM, Watkins JC. Diffusion of univalent ions across the lamellae of swollen phospholipids. Journal of Molecular Biology. 1965;**13**(1):238–252

[10] Vanlerberghe G. Niosomes and other non-phospholipid vesicular systems. In: Uchegbu IF, editor. Synthetic Surfactant Vesicles. Amsterdam: Harwood Academic Publishers; 2000. pp. 3–7

[11] Fendler JH. Membrane Mimetic Chemistry. New York, NY: Wiley and Sons; 1982

[12] Lasic DD. Liposomes. Physics to Applications. Amsterdam: Elsevier; 1993

[13] Lasic DD. Novel applications of liposomes. Trends in Biotechnology. 1998;**16**(7): 307–321

[14] Storm G, ten Kate MT, Working PK, Bakker-Woudenberg IA. Doxorubicin entrapped in sterically stabilized liposomes: Effects on bacterial blood clearance capacity of the mononuclear phagocyte system. Clinical Cancer Research: An Official Journal of the American Association for Cancer Research. 1998;**4**(1):111–115

[15] Torchilin VP. Recent advances with liposomes as pharmaceutical carriers. Nature Reviews Drug Discovery. 2005;**4**(2):145–160. DOI: 10.1038/nrd1632

[16] Gibbs BF, Kermasha S, Alli I, Mulligan CN. Encapsulation in the food industry: A review. International Journal of Food Science Nutrition. 1999;**50**(3):213–224

[17] Taylor TM, Davidson PM, Bruce BD, Weiss J. Liposomal nanocapsules in food science and agriculture. Critical Reviews in Food Science and Nutrition. 2005;**45**(7–8):587–605. DOI: 10.1080/10408390591001135

[18] Edwards KA, Baeumner AJ. Liposomes in analyses. Talanta. 2006;**68**(5):1421–1431

[19] Monnard PA, Deamer DW. Membrane self-assembly processes: Steps toward the first cellular life. The Anatomical Record. 2002;**268**(3):196–207. DOI: 10.1002/ar.10154

[20] Luisi PL, Ferri F, Stano P. Approaches to semi-synthetic minimal cells: a review. Naturwissenschaften. 2006;**93**(1):1–13. DOI: 10.1007/s00114-005-0056-z

[21] Uchegbu IF, Vyas SP. Non-ionic surfactant based vesicles (niosomes) in drug delivery. International Journal of Pharmaceutics. 1998;**172**(1–2):33–70. http://dx.doi.org/10.1016/S0378-5173(98)00169-0

[22] Uchegbu IF, Florence AT. Non-ionic surfactant vesicles (niosomes): Physical and pharmaceutical chemistry. Advances in Colloid and Interface Science. 1995;**581**(1):1–55

[23] Wang ZF, M. Analysis of sorbitan ester surfactants. 2. Capillary supercritical-fluid chromatography. High Resolution Chromatography. 1994;**17**:85–90

[24] Cottrell, T. van Peij, J. Sorbitan esters and polysorbates. In: Whitehurst RJ, editor. Emulsifiers in Food Technology. Oxford: Blackwell Publishing Ltd.; 2004. pp. 162–185

[25] Kato K, Walde P, Koine N, Ichikawa S, Ishikawa T, Nagahama R, Ishihara T, Tsujii T, Shudou M, Omokawa Y, Kuroiwa T. Temperature-sensitive nonionic vesicles prepared from Span 80 (sorbitan monooleate). Langmuir: The ACS Journal of Surfaces and Colloids. 2008;**24**(19):10762–10770. DOI: 10.1021/la801581f

[26] Hayashi K, Shimanouchi T, Kato K, Miyazaki T, Nakamura A, Umakoshi H. Span 80 vesicles have a more fluid, flexible and "wet" surface than phospholipid liposomes. Colloids and Surfaces B: Biointerfaces. 2011;**87**(1):28–35. DOI: 10.1016/j.colsurfb.2011.04.029

[27] Hayashi K, Tatsui T, Shimanouchi T, Umakoshi H. Membrane interaction between Span 80 vesicle and phospholipid vesicle (liposome): Span 80 vesicle can perturb and hemifuse with liposomal membrane. Colloids and Surfaces B: Biointerfaces. 2013;**106**:258–264. DOI: 10.1016/j.colsurfb.2012.12.022

[28] Omokawa Y, Miyazaki T, Walde P, Akiyama K, Sugahara T, Masuda S, Inada A, Ohnishi Y, Saeki T, Kato K. In vitro and in vivo anti-tumor effects of novel Span 80 vesicles containing immobilized Eucheuma serra agglutinin. International Journal of Pharmaceuticals. 2010;**389**(1–2):157–167. DOI: 10.1016/j.ijpharm.2010.01.033

[29] Brandner JD. The composition of NF-defined emulsifiers: Sorbitan monolaurate, monopalmitate, monostearate, monooleate, polysorbate 20, polysorbate 40, polysorbate 60, and polysorbate 80. Drug Development and Industrial Pharmacy. 1998;**24**(11):1049–1054. DOI: 10.3109/03639049809089948

[30] Lu DR, Rhodes D.G. Mixed compositions films of Spans and Tween-80 at the air-water interface. Langmuir: The ACS Journal of Surfaces and Colloids. 2000;**16**:8107–8112

[31] Yoshioka, T. Stermberg, B. Florence, A.T. Preparation and properties of vesicles (niosomes) of sorbitan monoesters (Span 20, 40, 60, and 80) and a sorbitan triester (Span 85). International Journal of Pharmaceuticals. 1994;**105**:1–6

[32] Kono K, Takagishi T. Temperature-sensitive liposomes. Methods in Enzymology. 2004;**387**:73–82. DOI: 10.1016/S0076-6879(04)87005-8

[33] Needham D, Dewhirst MW. The development and testing of a new temperature-sensitive drug delivery system for the treatment of solid tumors. Advanced Drug Delivery Reviews. 2001;**53**(3):285–305

[34] Nakata H, Miyazaki T, Iwasaki T, Nakamura A, Kidani T, Sakayama K, Masumoto J, Miura H. Development of tumor-specific caffeine-potentiated chemotherapy using a novel drug delivery system with Span 80 nano-vesicles. Oncological Reports. 2015;**33**(4):1593–1598. DOI: 10.3892/or.2015.3761

[35] Oberlin O, Fawaz O, Rey A, Niaudet P, Ridola V, Orbach D, Bergeron C, Defachelles AS, Gentet JC, Schmitt C, Rubie H, Munzer M, Plantaz D, Deville A, Minard V, Corradini N, Leverger G, de Vathaire F. Long-term evaluation of ifosfamide-related nephrotoxicity in children. Journal of Clinical Oncology. 2009;**27**(32):5350–5355. DOI: 10.1200/jco.2008.17.5257

[36] Hotz MA, Gong J, Traganos F, Darzynkiewicz Z. Flow cytometric detection of apoptosis: Comparison of the assays of in situ DNA degradation and chromatin changes. Cytometry. 1994;**15**(3):237–244. DOI: 10.1002/cyto.990150309

[37] Asai T, Ueda T, Itoh K, Yoshioka K, Aoki Y, Mori S, Yoshikawa H. Establishment and characterization of a murine osteosarcoma cell line (LM8) with high metastatic potential to the lung. International Journal of Cancer. 1998;**76**(3):418–422

[38] Elbayoumi TA, Torchilin VP. Enhanced cytotoxicity of monoclonal anticancer antibody 2C5-modified doxorubicin-loaded PEGylated liposomes against various tumor cell lines. European Journal of Pharmaceutical Science. 2007;**32**(3):159–168. DOI: 10.1016/j.ejps.2007.05.113

[39] Skinner R, Cotterill SJ, Stevens MC. Risk factors for nephrotoxicity after ifosfamide treatment in children: A UKCCSG Late Effects Group study. United Kingdom Children's Cancer Study Group. British Journal of Cancer. 2000;**82**(10):1636–1645. DOI: 10.1054/bjoc.2000.1214

[40] Patterson WP, Khojasteh A. Ifosfamide-induced renal tubular defects. Cancer. 1989;**63**(4):649–651

[41] Skinner R, Pearson AD, Coulthard MG, Skillen AW, Hodson AW, Goldfinch ME, Gibb I, Craft AW. Assessment of chemotherapy-associated nephrotoxicity in children with cancer. Cancer Chemotherapy and Pharmacology. 1991;**28**(2):81–92

[42] McCune JS, Friedman DL, Schuetze S, Blough D, Magbulos M, Hawkins DS. Influence of age upon ifosfamide-induced nephrotoxicity. Pediatric Blood Cancer. 2004;**42**(5):427–432. DOI: 10.1002/pbc.20011

[43] Wagner T. Ifosfamide clinical pharmacokinetics. Clinical Pharmacokinetics. 1994;**26**(6):439–456. DOI: 10.2165/00003088-199426060-00003

[44] Howell JE, Szabatura AH, Hatfield Seung A, Nesbit SA. Characterization of the occurrence of ifosfamide-induced neurotoxicity with concomitant aprepitant. Journal of Oncology Pharmacy Practice. 2008;**14**(3):157–162. DOI: 10.1177/1078155208093930

[45] Cheung MC, Jones RL, Judson I. Acute liver toxicity with ifosfamide in the treatment of sarcoma: A case report. Journal of Medical Case Reports. 2011;**5**:180. DOI: 10.1186/1752-1947-5-180

[46] Yonemoto T, Ishii T, Takeuchi Y, Hagiwara Y, Iwata S, Tatezaki S. Recently intensified chemotherapy for high-grade osteosarcoma may affect fertility in long-term male survivors. Anticancer Research. 2009;**29**(2):763–767

[47] Ridola V, Fawaz O, Aubier F, Bergeron C, de Vathaire F, Pichon F, Orbach D, Gentet JC, Schmitt C, Dufour C, Oberlin O. Testicular function of survivors of childhood cancer: A comparative study between ifosfamide- and cyclophosphamide-based regimens. European Journal of Cancer. 2009;**45**(5):814–818. DOI: 10.1016/j.ejca.2009.01.002

[48] Longhi A, Macchiagodena M, Vitali G, Bacci G. Fertility in male patients treated with neoadjuvant chemotherapy for osteosarcoma. Journal of Pediatric Hematology/Oncology. 2003;**25**(4):292–296

[49] Thomson B, Hawkins D, Felgenhauer J, Radich J. RT-PCR evaluation of peripheral blood, bone marrow and peripheral blood stem cells in children and adolescents undergoing VACIME chemotherapy for Ewing's sarcoma and alveolar rhabdomyosarcoma. Bone Marrow Transplantation. 1999;**24**(5):527–533. DOI: 10.1038/sj.bmt.1701939

[50] Lotz JP, Andre T, Donsimoni R, Firmin C, Bouleuc C, Bonnak H, Merad Z, Esteso A, Gerota J, Izrael V. High dose chemotherapy with ifosfamide, carboplatin, and etoposide combined with autologous bone marrow transplantation for the treatment of poor-prognosis germ cell tumors and metastatic trophoblastic disease in adults. Cancer. 1995;**75**(3):874–885

[51] Williams D, Crofton PM, Levitt G. Does ifosfamide affect gonadal function? Pediatric Blood Cancer. 2008;**50**(2):347–351. DOI: 10.1002/pbc.21323

[52] Hayashi K, Tatsui T, Shimanouchi T, Umakoshi H. Enhanced cytotoxicity for colon 26 cells using doxorubicin-loaded sorbitan monooleate (Span 80) vesicles. International Journal of Biological Science. 2013;**9**(2):142–148. DOI: 10.7150/ijbs.5453

[53] Iwamoto T. Clinical application of drug delivery systems in cancer chemotherapy: Review of the efficacy and side effects of approved drugs. Biological & Pharmaceutical Bulletin. 2013;**36**(5):715–718

[54] Eckes J, Schmah O, Siebers JW, Groh U, Zschiedrich S, Rautenberg B, Hasenburg A, Jansen M, Hug MJ, Winkler K, Putz G. Kinetic targeting of pegylated liposomal doxorubicin: A new approach to reduce toxicity during chemotherapy (CARL-trial). BMC Cancer. 2011;**11**:337. DOI: 10.1186/1471-2407-11-337

[55] Maeng JH, Lee DH, Jung KH, Bae YH, Park IS, Jeong S, Jeon YS, Shim CK, Kim W, Kim J, Lee J, Lee YM, Kim JH, Kim WH, Hong SS. Multifunctional doxorubicin loaded superparamagnetic iron oxide nanoparticles for chemotherapy and magnetic resonance imaging in liver cancer. Biomaterials. 2010;**31**(18):4995–5006. DOI: 10.1016/j.biomaterials.2010.02.068

[56] Kato S, Mori S, Kodama T. A novel treatment method for lymph node metastasis using a lymphatic drug delivery system with nano/microbubbles and ultrasound. Journal of Cancer. 2015;**6**(12):1282–1294. DOI: 10.7150/jca.13028

[57] Kato S, Shirai Y, Kanzaki H, Sakamoto M, Mori S, Kodama T. Delivery of molecules to the lymph node via lymphatic vessels using ultrasound and nano/microbubbles. Ultrasound in Medicine & Biology. 2015;**41**(5):1411–1421. DOI: 10.1016/j.ultrasmedbio.2014.12.014

[58] Sato T, Mori S, Arai Y, Kodama T. The combination of intralymphatic chemotherapy with ultrasound and nano-/microbubbles is efficient in the treatment of experimental tumors in mouse lymph nodes. Ultrasound in Medicine & Biology. 2014;**40**(6):1237–1249. DOI: 10.1016/j.ultrasmedbio.2013.12.012

Chapter 6

Influence of Exogenously Supplemented Caffeine on Cell Division, Germination, and Growth of Economically Important Plants

Wojciech Sledz, Agata Motyka, Sabina Zoledowska, Agnieszka Paczek, Emilia Los and Jacek Rischka

Additional information is available at the end of the chapter

http://dx.doi.org/10.5772/67799

Abstract

Caffeine is a plant secondary metabolite of antiherbivory, allelopathic, and antibacterial activity. In our previous study, caffeine was shown to be an effective agent toward plant pathogenic bacteria causing high economic losses in crop production worldwide. Current study indicated that growth media supplementation with soil or plant extract did not interfere with antibacterial action of caffeine against *Clavibacter michiganensis, Dickeya solani, Pectobacterium atrosepticum, Pectobacterium carotovorum, Pseudomonas syringae, Ralstonia solanacearum, and Xanthomonas campestris*. The impact of caffeine on plant cell division, seed germination and growth of economically important plants was evaluated to assess possible applicability of caffeine in plant protection field. Caffeine impaired plant cell division process and inhibited *in vitro* germination of tomato and lettuce. Regeneration of potato explants was also negatively affected by the addition of caffeine. However, caffeine spraying or watering of tomato, lettuce and cabbage grown in soil did not impair plant development. Although the tested plants accumulated caffeine, its inner quantity was reduced by peeling and/or cooking. According to the results, caffeine warrants additional attention as a useful, natural compound designated for the control of bacterial plant pathogens. Proposed treatment seems promising especially in the case of providing protection for overwinter-stored table potato tubers.

Keywords: antimicrobials, *Brassica oleracea*, *Clavibacter* sp., *Dickeya* sp., *Lactuca sativa*, *Pectobacterium* sp., plant protection, *Pseudomonas* sp., *Ralstonia* sp., *Solanum lycopersicum*, *Solanum tuberosum*, *Vicia faba*, *Xanthomonas* sp.

1. Introduction

Plants produce a broad range of secondary metabolites exhibiting antibiotic, antifungal, antiviral, antigerminative, allelopathic, UV light absorbing, insecticidal, or even antiherbivore activities [1]. Caffeine (1,3,7–trimethylxanthine) is one of over 12,000 alkaloids of plant origin [2]. So far, caffeine has captured attention for its pharmacological activity, being the most widely consumed psychoactive substance in the world [3]. But little is known about its potential application in plant protection. Until now, it was reported that caffeine could be used to eradicate or repel molluscs, insects, frogs, or birds [4–7]. Also, the antibacterial activity of caffeine toward microbes inhabiting different ecological niches was demonstrated. This substance impaired growth of human pathogens like *Escherichia coli* O157:H7 responsible for approximately 73,500 cases of foodborne illnesses per year [8], constituents of natural human microflora such as *Streptococcus mutans* [9], or terrestrial and aquatic inhabitants like *Pseudomonas fluorescens* [10]. To the best of our knowledge, caffeine bactericidal properties against plant pathogenic bacteria have been examined so far by a few groups only. Kim and Sano [11] inoculated transgenic tobacco plants producing 1.8 μg caffeine per gram of fresh weight with *Pseudomonas syringae* pv. *glycinea* and noted remarkably lower disease severity in comparison with the nontransgenic plants. As many problems arise with the approval of genetically modified organisms, scientific attention focused on exogenously applied caffeine. Caffeine direct bactericidal action against *P. syringae* pv. *glycinea* was correlated with the increasing concentration of this compound [11]. Subsequently, Sledz et al. [12] evaluated antibacterial activity of caffeine toward broad spectrum of plant pathogenic bacteria causing economic losses in crop and ornamental plant production worldwide: *Clavibacter michiganensis* subsp. *sepedonicus* (Cms), *Dickeya solani* (Dsol), *Pectobacterium atrosepticum* (Pba), *Pectobacterium carotovorum* subsp. *carotovorum* (Pcc), *Pseudomonas syringae* pv. *tomato* (Pst), *Ralstonia solanacearum* (Rsol), and *Xanthomonas campestris* pv. *campestris* (Xcc). Caffeine inhibited growth of the above-listed phytopathogens in broth cultures, increased their generation time, triggered morphological abbreviations, and finally exhibited bactericidal effect in concentrations from 40 to 100 mM [12]. Moreover, application of this compound reduced disease symptoms caused by Dsol on potato slices, whole potato tubers, and chicory leaves [12]. In addition, it was shown that the plant pathogenic bacterium tested could not develop any resistance to the caffeine treatment [12].

Taking into account these data, caffeine seems to be a promising antimicrobial agent that might be utilized in the plant protection field, especially because of the limited amount of possible alternatives [13]. In the past, worldwide spread of multidrug-resistant microorganisms suggested more prudent uses of antibiotics in agriculture [14], thus possible plant control approaches seem even more restrained nowadays. In general, mostly preventive procedures are implemented to reduce economic damage triggered by plant pathogenic bacteria in the field, transportation, or storage [13].

In this work, we undertook further studies on evaluating possible applicability of caffeine as an antimicrobial agent to be used in agriculture. We investigated whether caffeine retains its action against plant pathogenic bacteria Cms, Dsol, Pba, Pcc, Pst, Rsol, and Xcc in the presence of substances occurring in soil or plant extracts. Moreover, the impact of caffeine treatment on plants of economic importance was studied. The effect of caffeine supplementation on plant cell division was shown in the sprouts of broad bean. Furthermore, we evaluated the influence of caffeine on plant germination and growth both *in vitro* and in soil. Last but not least, caffeine accumulation in the tested plants was investigated. In addition, the effect of peeling and/or cooking on inner caffeine content in potato tubers was evaluated. Altogether, this study provides an insight into possible ways of exploiting antibacterial activity of caffeine in plant protection field.

2. Materials and methods

2.1. Bacterial strains

Strains of plant pathogenic bacteria used in this study are: *Clavibacter michiganensis* subsp. *sepedonicus* LMG 2889, *Dickeya solani* IFB 0099, *Pectobacterium atrosepticum* SCRI 1043, *Pectobacterium carotovorum* subsp. *carotovorum* SCRI 180, *Pseudomonas syringae* pv. *tomato* LMG 5093, *Ralstonia solanacearum* LMG 2294, and *Xanthomonas campestris* pv. *campestris* LMG 582.

2.2. Plant material

The following plants were used: lettuce (*Lactuca sativa* L. var. *capitata*, cv. Queen of May), tomato (*Solanum lycopersicum* L., cv. Baron, cv. Betalux), cabbage (*Brassica oleracea* L. *convar. capitata*, cv. First harvest), potato (*Solanum tuberosum* L., cv. Irga, or the breeding lines: LB-6 and LB-12 [15]), and broad bean (*Vicia faba* L., cv. Hangdown white).

2.3. Growth media and media with soil or plant extract supplementation

To prepare soil extract, 1000 g of Substral soil (Scotts, Warsaw, Poland) was mixed (30 min, 250 rpm) with 2000 ml of distilled water. The suspension was filtered through Whatman paper grade 1 (Sigma-Aldrich, St. Louis, USA), and the resulting filtrate was autoclaved for 30 min.

On the basis of soil extract, soil medium was prepared as follows: 200 ml of the soil extract was supplemented with 1 g of K_3PO_4, 2 g of NaCl, 0.5 g of NH_4NO_3, and 1 g of glucose. pH was adjusted to 7.2.

Potato, tomato, and cabbage extract media were prepared as listed here: 50 g of potato tubers, 7.5 g of tomato leaves, or 20 g of cabbage leaves were homogenized in 100 (potato) or 20 ml (cabbage and tomato) of Ringers buffer in extraction bags (Bioreba, Basel, Switzerland). In the case of potato tissue extract 0.02 g of diethyldithiocarbamic acid was added. Then the homogenates were centrifuged at 4000 rpm for 15 min. The supernatant was collected and supplemented with additional nutrients to culture Pst (0.1 g of glucose) or Rsol (0.1 g of glucose and 0.3 g of yeast extract). Plant extract media were sterilized in sequence with 5, 1.2, 0.8, 0.45, and 0.22 μm Minisart SRP Syringe Filters (Sartorius, Goettingen, Germany).

In vitro plant cultures were conducted on MS [16], ½ MS or basic plant growth medium (sucrose 30 g l^{-1} and agar 7.5 g l^{-1}) depending on the experiment.

2.4. Effect of soil and plant extracts on the antibacterial action of caffeine

To examine whether substances present in soil or plant extracts could impede antibacterial activity of caffeine toward plant pathogens, the growth of Cms, Dsol, Pba, Pcc, Pst, Rsol, and Xcc cultures in soil extract medium and plant extract medium containing 0, 5, 10, or 0, 1, 3 mM caffeine, respectively, was monitored for over 24 h by measuring the relative change in OD_{580}. Choice of the plant extract medium for testing the survival of a specific pathogen was based on the preferable host. Potato extract medium was used in the case of Cms, Dsol, Pba, Pcc, and Rsol. Pst was cultured in tomato extract medium, while Xcc was incubated in cabbage extract medium. The experiment was conducted in darkness at 28°C (the exception was Cms cultured at 21°C).

2.5. Effect of caffeine on plant cell division

Broad bean seeds were incubated in distilled water for 24 h (20 seeds per 200 ml) and then germinated on moistened Whatman paper at 20°C. The sprouts were transferred to Petri plates containing six layers of lignin and watered with 0 or 8 mM caffeine. The sprouts were then kept at 24°C for 72 h. After the sprouts were washed, their apical meristems were isolated and cut into 5-mm slices, which were fixed and stained according to the Feulgen protocol [17]. There were four control and four treated samples and one to two preparations per sample. Cells were observed with a light microscope at 500 to 1600× magnification. For each preparation, 1000 cells were selected at random and examined for exhibited mitotic phase and visible micronuclei as described by Evans et al. [18]. The following parameters were determined: the mean mitotic index (the percentage of dividing cells in the observed cell population), phase index (the percentage of cells in prophase, metaphase, anaphase, or telophase), and the frequency of micronuclei. The experiment was performed twice.

2.6. Effect of caffeine on seed germination

The seeds of lettuce (cv. Queen of May) and of tomato (cv. Betalux) were surface-sterilized in 5% $Ca(OCl)_2$, rinsed with sterile-distilled water, and then placed in Petri dishes (10 seeds per dish) on Whatman filter papers moistened with 5 ml of a caffeine solution at 0, 1, 3, 5, 8, 10, 15, or 20 mM. Each combination of seed type and caffeine concentration was represented by at least five Petri dishes, which were sealed with parafilm and kept at 24°C with a 16/8 h light/dark photoperiod. Germinated and non-germinated seeds were counted after 3 or 7 days in the case of lettuce and tomato, respectively. The experiment was performed twice.

2.7. Impact of caffeine on early growth of seedlings

The effect of caffeine on *in vitro* germination of cabbage (cv. First harvest) and tomato (cv. Baron and cv. Betalux) seeds and on the early growth of seedlings was evaluated by placing surface-sterilized seeds on basic plant growth medium supplemented with 0, 1, 5, or 8 mM caffeine.

After 10 days at 24°C with a 16/8 h light/dark photoperiod, germinated seedlings were transferred to ½ MS medium with the same caffeine concentrations as before. Plant growth and development were monitored for 1 month. Each combination of cabbage or tomato seeds and certain caffeine concentration treatment was represented by four plants, and the experiment was performed three times.

2.8. Effect of caffeine on explants regeneration

Explants of potato (LB-6 and LB-12) stem fragments were transferred to MS medium containing 0, 1, 5, or 8 mM caffeine, and their growth was monitored for 6 weeks at 24°C (16/8h light/dark photoperiod). After the experiment, plant heights were measured. Each combination of potato line and caffeine concentration was represented by three replicates. The experiment was performed three times.

2.9. Effect of caffeine spraying and watering on soil-grown plants

The spraying experiment included cabbage (cv. First harvest), lettuce (cv. Queen of May), and tomato (cv. Betalux). The seeds were germinated on moistened Whatman paper, and after 2 weeks the seedlings were planted in pots (27 × 31 × 4 cm) containing autoclaved soil. There were five rows of 10 plants per pot. The pots were kept at 20°C with a 16/8 h light/dark photoperiod and were watered every 3 days. After the seedlings had been grown in the pots for 10 days, they were sprayed (10 ml per pot) with an aqueous solution containing 0, 1, 5, or 8 mM caffeine. The caffeine was applied seven times over 6 weeks before plant heights were measured. The experiment was performed three times.

The effect of watering with caffeine was assessed on tomato (cv. Betalux) and lettuce (cv. Queen of May). Seeds were planted in pots (7 × 7 × 12 cm; nine seeds per pot and four pots per plant type) containing autoclaved soil. The plants were kept at 20°C with a 16/8 h light/dark photoperiod. Each pot was watered every 3 days with 30 ml of caffeine solution (0, 1, 5, or 8 mM). After 4 weeks, plant heights were measured. The experiment was performed twice.

2.10. Caffeine accumulation in plant tissue

Plants collected from experiments concerning the effects of caffeine on plant germination, growth, and development were examined for caffeine accumulation. Plant material was frozen and stored at −20°C. Later on, it was gently thawed, washed twice with distilled water, dried, and weighed. Afterwards, the tissue was frozen in liquid nitrogen and crushed into small pieces. Two extracts per sample were prepared in 1 ml of Milli-Q H_2O by heating two-thirds of the sample to 100°C for 20 min. A third extract was obtained by keeping one-third sample at 25°C for 20 min. All three extracts were pooled and filtered via a 0.45-μm Minisart SRP Syringe Filter (Sartorius). The filtrates were kept at 4°C before they were processed by high performance liquid chromatography (HPLC) with a Series 200 system (Perkin Elmer, Waltham, USA) and a C18 column (Sigma-Aldrich). A 15% methanol solution was used as a mobile phase. The retention time of caffeine was about 8.3 min.

Caffeine content in the samples was determined by calculating the surface area under the 280 nm absorbance peak in comparison to a standard curve obtained with different caffeine concentrations.

2.11. Impact of peeling and/or cooking on caffeine content

Five potato tubers (cv. Irga) were incubated in a 100 mM caffeine solution for 24 h at room temperature. Samples (0.8 g each) were collected from the peels and from the transitional and middle zones. Potato middle zone was a cube of approximately 3 × 3 × 3 cm originating from the center of the inner mass. Transitional zone enclosed between the middle zone and the peel. Caffeine was extracted from the potato tuber samples with dichloromethane. Caffeine content was assessed in the zones by gas chromatography (Clarus 600, Perkin Elmer) and the quantity of caffeine per gram of dry weight was subsequently calculated.

Other five potato tubers were incubated at room temperature in a 100 mM caffeine solution for 24 h and then cooked, with or without the peels, at 100°C for 20 min. Samples were collected and the caffeine content in specific zones was evaluated as described above.

2.12. Statistical analysis

Statistical significance within plant germination experiments was evaluated by Kruskal-Wallis test, followed by Dunn's test, while the impact of caffeine spraying and watering on plant heights was assessed with the Tukey's test (HSD). $p < 0.05$ was utilized.

3. Results and discussion

3.1. Effect of soil and plant extract on the antibacterial activity of caffeine

The growth dynamics of Dsol, Cms, Pba, Pcc, Pst, Rsol, and Xcc in soil extract media and plant extract media supplemented with 0, 5, 10 and 0, 1, 3 mM caffeine, respectively, was evaluated. Caffeine still reduced bacterial growth in such conditions (**Figure 1**). Interestingly, caffeine was the most effective against Xcc, Rsol, and Cms both in the case of soil and plant extract supplemented media (**Figure 1**). Observed inhibition pattern for all the tested pathogens was similar to the one reported by Sledz et al. [12]. It needs to be taken into account that the examined plant or soil extracts were autoclaved prior to use, and the metabolic activity of soil and plant microflora has also an impact on vastness and diversity of substances naturally occurring in the environment. Further research is needed to exclude possible inactivation of caffeine via complex formation with polyphenols or sequestration into chlorogenic acid complex [19]. Likewise, the impact of species capable of caffeine degradation, e.g. *Pseudomonas cepacia, Pseudomonas putida*, and *Serratia marcescens*, needs to be taken into consideration. In addition, the metabolites present in the

http://dx.doi.org/10.5772/67799

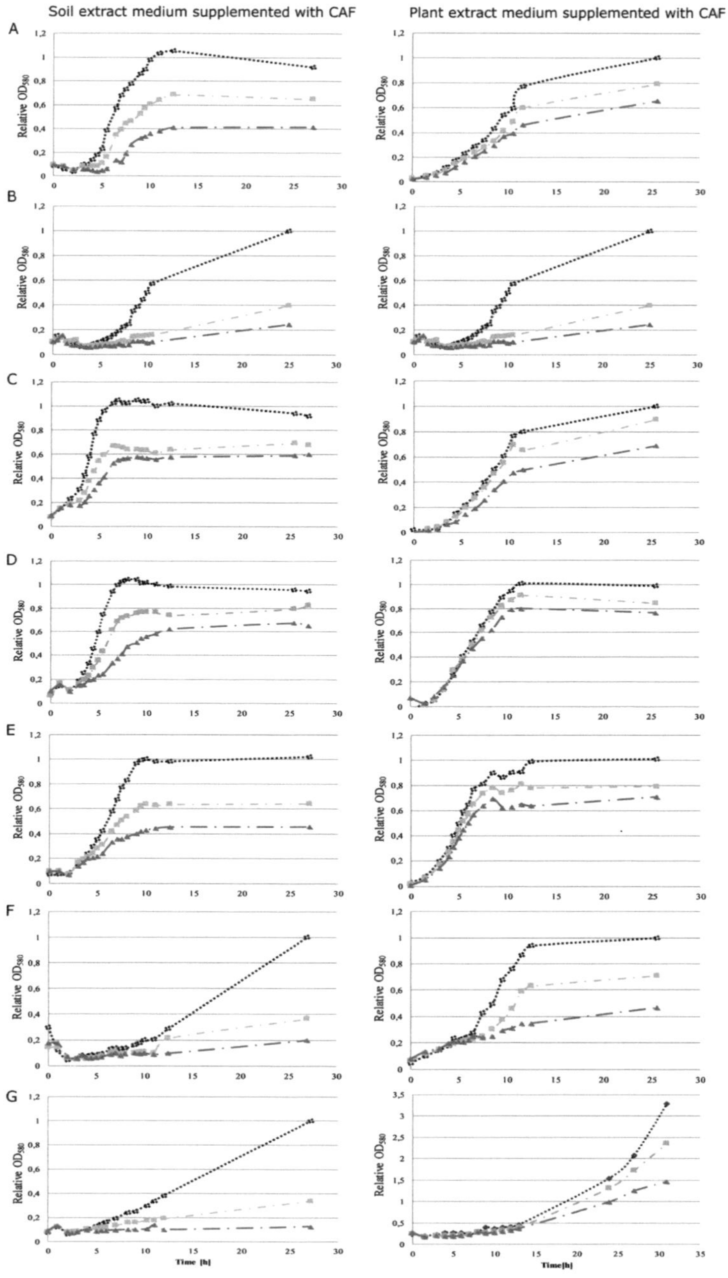

Figure 1. Influence of soil and plant extract on the antibacterial activity of caffeine against the following phytopathogens: (A) *Dickeya solani*, (B) *Ralstonia solanacearum*, (C) *Pectobacterium atrosepticum*, (D) *Pectobacterium carotovorum* subsp. *carotovorum*, (E) *Pseudomonas syringae* pv. *tomato*, (F) *Xanthomonas campestris* pv. *campestris*, and (G) *Clavibacter michiganensis* subsp. *sepedonicus*.

implemented plant and soil extracts are in more oxidized state than those enclosed inside plant tissue.

3.2. Influence of caffeine on plant cell division

To evaluate the effect of exogenous caffeine supplementation on plant cell division, we used *Vicia faba* L. model, widely utilized in studies on environmental mutagens [20]. The mitotic index in broad bean apical meristems was increased by 8 mM caffeine treatment in comparison with the non-treated controls (**Table 1**). This resulted mainly from higher percentage of cells undergoing the prophase state. Lack of caffeine treatment resulted in higher percentage of cells in later stages of cell division process, namely, metaphase, anaphase, and telophase. Besides, micronuclei were observed more frequently in the caffeine-treated cells than in the non-treated samples. Altogether, our results indicate that caffeine treatment resulted in higher frequency of cells undergoing earlier phases of cell division and having visible micronuclei, which points into symptoms of genome instability. Premature chromosome condensation resulting in apoptosis-like programmed cell death was postulated by Rybaczek et al., while investigating caffeine action on root meristems of *Vicia faba* [21]. Interestingly, Friedman and Waller [22] reported repression of mitosis and cell plate formation in coffee seeds exposed to 10 mM caffeine, while Valster and Hepler [23] observed that caffeine allows initiation of the cell plate formation but prevents its completion in living *Tradescantia* stamen hair cells. According to Valster and Hepler, the cytokinesis is affected by the inhibition of cytoskeletal torus formation during phragmoplast expansion [23].

Cell division parameter	Caffeine concentration	
	0 mM	**8 mM**
Mitotic index (%)	7.27 ± 0.5	9.53 ± 0.34
Prophase index (%) 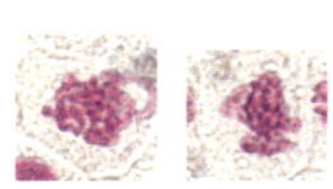	43.83 ± 8.67	68.57 ± 2.76
Metaphase index (%) 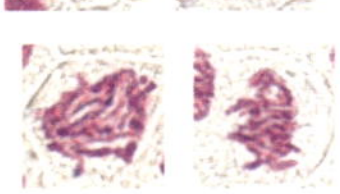	35.19 ± 6.46	26.69 ± 5.98
Anaphase index (%) 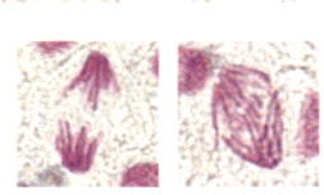	11.40 ± 0.47	2.97 ± 3.43
Telophase index (%) 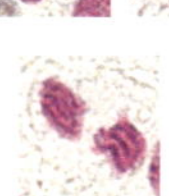	9.57 ± 1.91	1.77 ± 2.28
Micronuclei frequency (%) 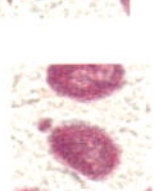	0.39 ± 0.12	2.71 ± 1.18

~3800 and ~1600 cells were analyzed for 0 and 8 mM caffeine treatment, respectively. Presented values are means ± SD.

Table 1. Influence of exogenous caffeine application on broad bean cell divisions.

http://dx.doi.org/10.5772/67799

3.3. Effect of caffeine on seed germination and plant development

In order to evaluate possible ways of applying caffeine against bacterial phytopathogens, the effect of caffeine on seed germination and early plant development *in vitro* was assessed. Caffeine reduced the germination rate of lettuce and tomato seeds on caffeine-moistened Whatman paper in a dose-dependent manner (**Figure 2**). Application of caffeine in concentrations higher than 5 and 8 mM significantly reduced the germination rate of lettuce and tomato, respectively, in comparison with the non-treated controls. This observation corresponds with studies on coffee seeds, as caffeine released from fallen, decomposing leaves of mature trees was proven to inhibit seed germination in the coffee plantations [22]. On the contrary, Avery et al. [7] found that caffeine did not reduce the germination of rice seeds under field conditions. We attribute

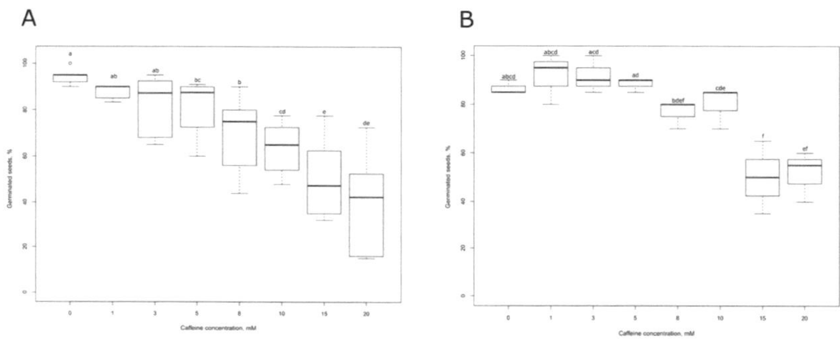

Figure 2. Effect of caffeine on seed germination of (A) lettuce cv. Queen of May and (B) tomato cv. Betalux. Germinated lettuce and tomato seeds were counted after 3 or 7 days of incubation, respectively.

	Cabbage cv. First harvest				Tomato cv. Baron				Tomato cv. Betalux			
Caffeine (mM)	0	1	5	8	0	1	5	8	0	1	5	8
14 days	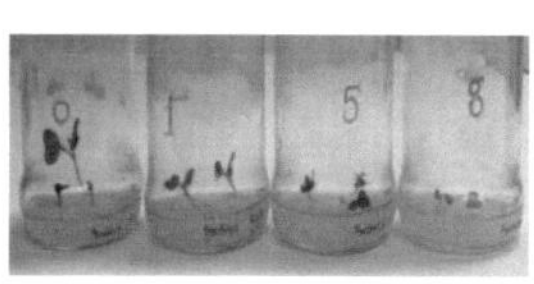											
Growth and development	1U	2U	3P	4DW	1U	2U	5	5	1U	2U	3DW	4W
30 days									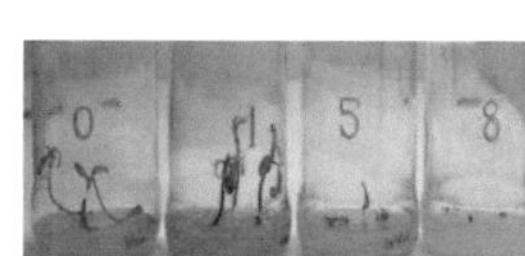			
Growth and development	1U	2DP	3DW	4W	1U	2DP	5	5	1 U	1DS	3DW	4W

Growth rate grading: 1—normal, 2—slower, 3—slow, 4—no growth, 5—no sprouted seeds. Development grading: U—uniform plant growth, D—darkening of the leaves, W—wilting, P—weaker plants, S—shed leaves. Photographs show growth and development of the representative plant for each treatment.

Table 2. Impact of caffeine on plant germination, growth and development *in vitro*.

this conclusion to ample water conditions required for rice cultivation that diminished the local concentration of caffeine.

Caffeine also impeded the germination, subsequent growth, and development of cabbage and tomato plants cultured on ½ MS medium (**Table 2**). Supplementation of the medium with caffeine in higher concentrations than 5 mM resulted in complete growth impairment of the tested plants (**Table 2**). In the case of plants growing on the 1 mM caffeine-enriched medium, they were weaker, exhibited slower growth rate, and certain darkening of the leaves after 30 days of incubation.

3.4. Effect of caffeine treatment on *in vitro*-grown and soil-grown plants

MS medium containing caffeine at concentrations higher than 5 mM completely inhibited *in vitro* regeneration of potato explants (cv. LB-6 and LB-12) (**Table 3**). Even application of 1 mM caffeine resulted in shorter potato plants of 5.3 ± 2.5 cm, in comparison with 10.7 ± 2.9 cm high controls. Similar pattern was shown in research on *Oryza sativa* L. by Smyth [24] who reported 2.5 mM caffeine suppression of shoot elongation by 50% and root elongation by 90%. Also in mung bean (*Phaseolus aureus*), Batish et al. [25] reported that caffeine reduced root number and length produced by hypocotyl cuttings.

Contrarily, spraying with 0, 1, 5, or 8 mM caffeine cabbage, lettuce, and tomato plants grown in soil did not significantly affect their growth or development as expressed by the plant heights measured after 6 weeks post planting (**Table 3**). Likewise, watering with 0, 1, 5, and 8 mM caffeine of lettuce and tomato plants grown in soil did not affect their heights that were

Plant	**Plants heights (cm)**			
	Caffeine concentration (mM)			
	0	**1**	**5**	**8**
Potato[1]	10.7 ± 2.9	5.3 ± 2.5	NG	NG
Tomato[2]	3.71 ± 0.62	3.33 ± 0.59	3.38 ± 0.53	3.26 ± 0.75
Cabbage[2]	6.56 ± 1.07	6.47 ± 1.07	6.32 ± 1.47	6.75 ± 1.13
Lettuce[2]	8.04 ± 1.40	8.58 ± 2.00	8.00 ± 1.21	8.36 ± 1.97
Tomato[3]	10.16 ± 1.53	10.23 ± 1.87	8.94 ± 1.83	8.46 ± 0.95
Lettuce[3]	10.35 ± 0.68	10.77 ± 1.50	9.88 ± 0.82	10.18 ± 1.00

[1]Micropropagation: Plants were grown on MS medium containing caffeine. Plants heights were measured after 6 weeks of incubation at 24°C (16/8 h light/dark photoperiod).

[2]Spraying: The seeds were germinated on moistened Whatman paper. After 2 weeks, they were planted in pots with autoclaved soil. Plants were grown at 20°C (16/8 h light/dark photoperiod) and were watered every 3 days. After 10 days, they were sprayed (10 ml per pot) with an aqueous solution containing caffeine. The caffeine was applied seven times over 6 weeks before the plant heights were measured.

[3]Watering: Seeds were planted in autoclaved soil. The plants were grown at 20°C (16/8 h light/dark photoperiod). Each pot was watered every 3 days with 30 ml of caffeine solution. After 4 weeks, plant heights were measured.

NG—no growth. Values are means ± SD.

Table 3. Effect of caffeine treatment on the heights of potato, tomato, cabbage, and lettuce plants.

measured after 4 weeks of continuous growth (**Table 3**). This corresponds with Hollingsworth et al. [4] stating that 2% caffeine caused no phytotoxicity symptoms when it was sprayed on four varieties of 4-week-old lettuce plants growing in the greenhouse. They also observed no lesions on leaves or roots of any of the oncidium orchids. The only serious symptoms like yellowing of the leaves followed by necrosis appeared after several days on excised leaves of lettuce and cabbage after being dipped in caffeine solutions ranging from 0.5 to 2.0% [4].

3.5. Caffeine accumulation in plant tissue

HPLC analysis revealed that caffeine is accumulated in plants that have been treated with this compound (**Table 4**). The accumulation of caffeine was much greater if the plants had been exposed to caffeine on Whatman paper or on MS medium rather than in soil (**Table 4**). Interestingly, the amount of caffeine accumulated in tomato leaf tissue was much higher than in the stem or root tissues. Contrarily, lettuce leaves did not exhibit higher caffeine accumulation level than the corresponding sprouts (**Table 4**). In conclusion, the level of caffeine accumulation depends strongly on caffeine application method and varied between the investigated plant organs. The latter observation corresponds with unequal distribution of caffeine within plant species capable of synthesizing caffeine. For example, *Camellia sinensis* var. *sinensis* contains 2.8% caffeine in its foliage, while *Coffea arabica* seedlings contain caffeine mainly in the leaves and cotyledons at concentrations ranging from 0.8 to 1.9%. Caffeine is absent, however, in roots and in older, brown parts of *C. arabica* shoots [26]. Besides, an interesting observation was reported by Bustos [27] that stated caffeine accumulation in aromatic herbs like sage or oregano when they were intercropped with coffee.

3.6. Impact of peeling and/or cooking on caffeine accumulation

The concentration of caffeine in dry potato tissue was determined after tubers were incubated in a 100 mM caffeine solution at room temperature without subsequent cooking or with the

Plant	Plant organ	Caffeine concentration in plant tissue (mg g^{-1})			
		Caffeine concentration in the medium			
		0 mM	1 mM	5 mM	8 mM
Tomato	Leaves[1]	0.0559	0.0756	0.3467	0.3906
	Stems[1]	0.0072	0.0069	0.0613	0.0214
	Roots[1]	0.0040	0.0666	0.0786	0.0334
Lettuce	Sprouts[2]	0.0001	0.2651	1.8485	2.9674
	Leaves[1]	0.0354	0.0188	0.1580	0.0708
Potato	Plants[3]	0.1865	1.1396	3.9577	2.5106

[1]Plants grown in soil at 20°C (16/8 h light/dark photoperiod).
[2]Seeds sprouted on Whatman paper at 24°C (16/8 h light/dark photoperiod).
[3]Plants grown *in vitro* on MS medium at 24°C (16/8 h light/dark photoperiod).

Table 4. Influence of application method on the accumulation of caffeine in plant tissue.

cooking (±prior peeling) at 100°C. The caffeine concentration in uncooked potatoes was higher in the peel than in the middle or transitional zone of the tuber (**Figure 3**). Also, subjecting potatoes to cooking significantly reduced the overall caffeine content in the tuber tissue. We observed that total caffeine concentration was the lowest when potatoes were peeled before cooking (**Figure 3**). Importantly, *Solanum tuberosum* L. cv. Irga was used in this study, but we suspect differences in effectiveness of caffeine washing during cooking between potato cultivars, because their pectins vary in branching, methylation, and acetylation level, which can have an effect on potent caffeine removal [28].

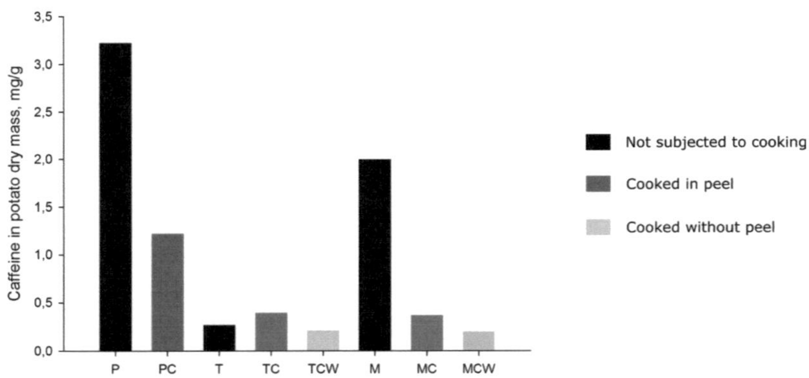

Figure 3. Caffeine concentration in the tissue of caffeine-treated potato tubers. Caffeine accumulated in the tissue originating from the following tuber zone after the indicated treatment: P—peel, without cooking; PC—peel, after cooking; T—transitional zone, without cooking; TC—transitional zone, after cooking; TCW—transitional zone, cooked without the peel; M—middle zone, without cooking; MC—middle zone, after cooking; MCW—middle zone, cooked without the peel.

4. Conclusions

World population is growing with an annual rate of 1.2%, meaning 77 million people per year [29], thus providing for food security and its safety appears crucial nowadays. Caffeine seems to be an attractive alternative for crop protection as it eradicates or repels molluscs, insects, frogs, birds, and phytopathogens [4–7, 12]. Even in the presence of compounds appearing in soil or plant extracts caffeine retained its inhibitory effect against Dsol, Pba, Pcc, Pst, Rsol, and Xcc, all mentioned by Mansfield et al. [30] in the list of top 10 plant pathogenic bacteria based on scientific and/or economic importance. So far, little is known about the possible ways to apply caffeine in agriculture. By now we have demonstrated that caffeine implementation on crop seeds could interfere with plant cell division and might inhibit the germination process. Thus, caffeine may be implemented before placing the potato seeds in storage, where inhibition of germination is an additional advantage. Importantly, watering and spraying of sprouts and the whole plants were proven not to interfere with further plant growth and development, so could be applied to agriculture in this form. Furthermore, our results showed that caffeine accumulated mainly in the peel of potato tuber and cooking significantly reduced the final caffeine content in all the tuber zones (especially while potatoes were peeled prior to thermal treatment).

As caffeine is obtained in commercial quantities by synthesis or as a by-product of the decaffeination process, the cost of the proposed treatment would not be high. Avery et al. calculated that rice treatment with 1% caffeine would cost the producers about 4$ ha^{-1} [7]. Not without importance is the fact that caffeine is readily soluble in water, which prevents its environmental accumulation. Moreover, caffeine is a common food additive of generally regarded as safe (GRAS) status, ingested directly in beverages such as tea or coffee throughout the world and even now it remains the fourth most frequently detected organic wastewater contaminant in the U.S. streams [31].

In conclusion, we think caffeine as a natural compound could be implemented effectively in agriculture in order to protect economically important crops and ornamentals from plant pathogenic bacteria.

Acknowledgements

This study was funded by the National Science Center in Poland via grant number NN 310150535 (MNiSzW 1505/B/P01/2008/35) dedicated to WS. We are also very grateful to Professor Ewa Łojkowska for suggestions, fruitful discussions, and everlasting support.

Author details

Wojciech Sledz*, Agata Motyka, Sabina Zoledowska, Agnieszka Paczek, Emilia Los and Jacek Rischka

*Address all correspondence to: sledz@biotech.ug.edu.pl

Department of Biotechnology, Intercollegiate Faculty of Biotechnology, University of Gdansk and Medical University of Gdansk, Gdansk, Poland

References

[1] Bourgaud F, Gravot A, Milesi S, Gontier E. Production of plant secondary metabolites: a historical perspective. Plant Sci 2001;161:839-51. doi:10.1016/S0168-9452(01)00490-3.

[2] Croteau R, Kutchan T. Natural products (secondary metabolites). Rockville: American Society of Plant Biologists; 2000. Available from: http://www.science.lecture.ub.ac.id/files/2012/04/plant-biosynthesis1.pdf

[3] Gilbert RM. Caffeine consumption. Prog Clin Biol Res 1984;158:185-213.

[4] Hollingsworth RG, Armstrong JW, Campbell E. Caffeine as a novel toxicant for slugs and snails. Ann Appl Biol 2003;142:91-7. doi:10.1111/j.1744-7348.2003.tb00233.x.

[5] Nathanson J. Caffeine and related methylxanthines: possible naturally occurring pesticides. Science (80) 1984;226:184-7. http://dx.doi.org/10.1126/science.6207592

[6] Campbell EW, Kraus F. Neotropical frogs in Hawaii: status and management options for an unusual introduced pest. USDA Natl Wildl Res Cent – Staff Publ 2002. Available from: http://digitalcommons.unl.edu/icwdm_usdanwrc/469

[7] Avery ML, Werner SJ, Cummings JL, Humphrey JS, Milleson MP, Carlson JC, et al. Caffeine for reducing bird damage to newly seeded rice. Crop Prot 2005;24:651-7. doi:10.1016/j.cropro.2004.11.009.

[8] Ibrahim SA, Salameh MM, Phetsomphou S, Yang H, Seo CW. Application of caffeine, 1,3,7-trimethylxanthine, to control *Escherichia coli* O157:H7. Food Chem 2006;99:645-50. doi:10.1016/j.foodchem.2005.08.026.

[9] Almeida AAP, Naghetini CC, Santos VR, Antonio AG, Farah A, Glória MBA. Influence of natural coffee compounds, coffee extracts and increased levels of caffeine on the inhibition of *Streptococcus mutans*. Food Res Int 2012;49. doi:10.1016/j.foodres.2012.07.026.

[10] Ramanavičienė A, Mostovojus V, Bachmatova I. Anti-bacterial effect of caffeine on *Escherichia coli* and *Pseudomonas fluorescens*. Acta Medica Cordoba 2003. Available from: http://www.elibrary.lt/resursai/LMA/Acta%20medica%20Lituanica/A-185.pdf

[11] Kim Y-S, Sano H. Pathogen resistance of transgenic tobacco plants producing caffeine. Phytochemistry 2008;69:882-8. doi:10.1016/j.phytochem.2007.10.021.

[12] Sledz W, Los E, Paczek A, Rischka J. Antibacterial activity of caffeine against plant pathogenic bacteria. Acta Biochim 2015;62:605-12. http://dx.doi.org/10.18388/abp.2015_1092

[13] Czajkowski R, Pérombelon MCM, van Veen JA, van der Wolf JM. Control of blackleg and tuber soft rot of potato caused by *Pectobacterium* and *Dickeya* species: a review. Plant Pathol 2011;60:999-1013. doi:10.1111/j.1365-3059.2011.02470.x.

[14] Lipsitch M, Singer RS, Levin BR. Antibiotics in agriculture: when is it time to close the barn door? Proc Natl Acad Sci U S A 2002;99:5752-4. doi:10.1073/pnas.092142499.

[15] Austin S, Lojkowska E, Ehlenfeldt K, Kelman A. Fertile interspecific somatic hybrids of *Solanum*: a novel source of resistance to *Erwinia* soft rot. Phytopathology 1988;78: 1216-20. Available from: https://www.apsnet.org/publications/phytopathology/backissues/Documents/1988Articles/Phyto78n09_1216.pdf

[16] Murashige T, Skoog F. A revised medium for rapid growth and bio assays with tobacco tissue cultures. Physiol Plant 1962; 15 doi: 10.1111/j.1399-3054.1962.tb08052.x

[17] Feulgen R, Rossenbeck H. Microchemical test for nucleic acid of the thymonucleic acid type and the selective staining of cell nuclei in microscopic preparations. Z Physiol Chem 1924;135:203.

[18] Evans HJ, Keary GJ, Tonkinson SM. The use of colchicine as an indicator of mitotic rate in broad bean root meristems. J Genet 1957;55:487-502. doi:10.1007/BF02984066.

[19] Ashihara H, Crozier A. Caffeine: a well known but little mentioned compound in plant science. Trends Plant Sci 2001;6:407-13. Available from: http://dx.doi.org/10.1016/S1360-1385(01)02055-6.

[20] Qian X. Mutagenic effects of chromium trioxide on root tip cells of Vicia faba. J Zhejiang Univ A 2004;5:1570-6. doi:10.1631/jzus.2004.1570.

[21] Rybaczek D, Musiałek MW, Balcerczyk A, Lockshin R, Zakeri Z, Mea M, et al. Caffeine-induced premature chromosome condensation results in the apoptosis-like programmed cell death in root meristems of Vicia faba. PLoS One 2015;10:e0142307. doi:10.1371/journal.pone.0142307.

[22] Friedman J, Waller GR. Caffeine hazards and their prevention in germinating seeds of coffee (Coffea arabica L.). J Chem Ecol 1983;9:1099-106. doi:10.1007/BF00982214.

[23] Valster A, Hepler P. Caffeine inhibition of cytokinesis: effect on the phragmoplast cytoskeleton in living *Tradescantia* stamen hair cells. Protoplasma 1997; 196 DOI: 10.1007/BF01279564

[24] Smyth DA. Effect of methylxanthine treatment on rice seedling growth. J Plant Growth Regul 1992;11:125-8. doi:10.1007/BF00198025.

[25] Batish DR, Singh HP, Kaur M, Kohli RK, Yadav SS. Caffeine affects adventitious rooting and causes biochemical changes in the hypocotyl cuttings of mung bean (Phaseolus aureus Roxb.). Acta Physiol Plant 2008;30:401-5. doi:10.1007/s11738-007-0132-4.

[26] Ashihara H, Sano H, Crozier A. Caffeine and related purine alkaloids: biosynthesis, catabolism, function and genetic engineering. Phytochemistry 2008;69:841-56. doi:10.1016/j.phytochem.2007.10.029.

[27] Bustos AP. Allelochemical effects of aromatic species intercropped with coffee (Coffea arabica L.) in Puebla, Mexico 2007. Available from: http://citeseerx.ist.psu.edu/viewdoc/download?doi=10.1.1.600.6402&rep=rep1&type=pdf

[28] van Marle JT, Stolle-Smits T, Donkers J, van Dijk C, Alphons GJ, Voragen A, et al. Chemical and microscopic characterization of potato (*Solanum tuberosum* L.) cell walls during cooking. J Agr Food Chem 1997;45. doi:10.1021/JF960085G.

[29] Carvalho FP. Agriculture, pesticides, food security and food safety. Environ Sci Policy 2006;9:685-92. doi:10.1016/j.envsci.2006.08.002.

[30] Mansfield J, Genin S, Magori S, Citovsky V, Sriariyanum M, Ronald P, et al. Top 10 plant pathogenic bacteria in molecular plant pathology. Mol Plant Pathol 2012;13:614-29. doi:10.1111/j.1364-3703.2012.00804.x.

[31] Kolpin D, Furlong E, Meyer M. Pharmaceuticals, hormones, and other organic wastewater contaminants in US streams, 1999–2000: a national reconnaissance. Sci Technol 2002;36 doi: 10.1021/es011055j

Chapter 7

Chemistry and Biotransformation of Coffee By-Products to Biofuels

Bianca Yadira Pérez-Sariñana and Sergio Saldaña-Trinidad

Additional information is available at the end of the chapter

http://dx.doi.org/10.5772/intechopen.68598

Abstract

Coffee is one of the most consumed infusion drinks in the world and contains a large variety of chemical compounds responsible for their sensory qualities and their effects on the body. The beneficial effects of coffee have been attributed only to its most important and researched ingredient, caffeine, but now it is known that other components have also contributed to its properties. Due to a huge demand for this product, large amounts of waste are generated in the coffee industry, which are toxic and represent serious environmental problems. During the process of mechanical extraction of the coffee seed, residues generated are: pulp, mucilage and parchment, mainly. Coffee cherry consists of soluble carbohydrates, insoluble polysaccharides, lipids, nitrogenous components, caffeine and minerals. More than 50% is considered a waste; it no longer has any commercial application, knowing that its components could be exploited for the production of inputs and energy. This chapter presents the chemistry and biotransformation of by-products and coffee residues into second-generation biofuels, which can be bioethanol, biogas and biodiesel by fermentation, anaerobic digestion and trans-esterification, respectively. Biofuels offer greater energy security, lower emissions of greenhouse gases and particulate matter, rural development, reduced demand for oil, among others.

Keywords: coffee cherry components, green coffee, roasted coffee, caffeine, biotransformation

1. Introduction

1.1. Coffee cherry

The coffee is the fruit and dried seed of the coffee plant regardless of whether it has been toasted or ground of the plant of the genus *Coffea* (**Figure 1**), generally of the cultivated species, and the products of these in their different stages of the process and use intended for human consumption. The fruits of coffee are often called cherries, since their appearance is like these and each one contains two hemispherical seeds. Coffee cherry is the set of non-dried fruits of *Coffea* plants after harvest [1].

According to the International Coffee Organization (ICO), in the period 2015–2016, countries around the world produced 147,997 in thousands of bags of 60 kg of coffee, of which two main varieties are produced: arabica and robusta. On the other hand, the consumption of coffee was 151,303 in thousands of bags of 60 kg of coffee. Data of global coffee production and consumption are summarized in **Table 1**.

Mexico contributed more than 2800 in thousands of bags of 60 kg, the most of the production is distributed in 13 states, including Chiapas, Veracruz, Puebla and Oaxaca, where more than 80% of production is concentrated. Approximately 402,099 tons of coffee cherry was grown in Chiapas, giving it the first place in national production; due to this, Mexico is the ninth largest coffee producer in the world [2].

In the historic part, coffee arrived in Mexico more than 200 years ago, entering Veracruz from Cuba. During the time of 1876–1911 (the Government of Porfirio Díaz), coffee plantation grew importantly in large specialized farms and later became an activity of small producers, mostly of indigenous origin.

A total of 97% of the coffee produced in Mexico is under the shade of trees, which respects the balance of the environment and protects many varieties of plants and animals. The strong harvest season covers the months from October to March.

Due to the characteristics of Mexican soils where coffee grows, mostly of volcanic type, its flavour is very characteristic, and its aroma is intense and with notes of chocolate, spices and flowers. This has served the Mexican coffee to receive two appellations of origin: Veracruz coffee and Chiapas coffee.

1.2. Biofuels—characteristics and advantages

Most economic scenarios are based on growth in global energy demand over the next 20 years. In this sense, nuclear and renewable energy as biomass, wind, hydropower, solar, photovoltaic, geothermal, etc., although do expansion, will remain secondary compared to fossil fuels, rising from 23.7 to 30% by 2040, concentrated in transportation sector and the oil industry mainly [3, 4].

In this context, the use of biofuels (fuels of biological origin) has huge potential and has a stronger expansion compared to other renewable alternatives. Biofuels are produced from biomass, a renewable resource provided that the crop cycle is respected [5].

Figure 1. Coffee cherry fruit (source: Authors).

Crop year (2015–2016)	Production	%total	Consumption	%total
Africa	16,831	11.37	10,815	7.15
Asia & Oceania	47,428	32.05	31,609	20.89
Mexico & Central America	16,739	11.31	5257	3.47
South America	66,997	45.27	24,717	16.34
Europe			50,870	33.62
North America			28,035	18.53

Table 1. Data of global coffee production and consumption.

Biofuels offer several advantages. It is considered that by reducing the demand for fossil fuels, biofuels could make the energy supply safer. Its use would also reduce import costs to countries with energy deficits and provide better trade balance and balance of payments. The emissions of greenhouse gas, carbon monoxide and particulate matter can be significantly reduced. Biofuels can also improve vehicle performance; in fact, the lubricity of biodiesel prolongs the life of conventional diesel engines [6].

Business will be generated and an increase in economic activity will be allowed with the transition to biofuels. Biofuels are renewable and both bioethanol and biodiesel are clean combustion. Another important aspect is that they can be marketed easier than other alternatives, because they can be stored and distributed using existing infrastructure. Biofuels should play a significant role in climate change policies and this will certainly open up opportunities for the development of biofuels in developing countries [7].

2. Main varieties of coffee

Robusta coffee (*Coffea canephora*) comes from Central Africa, which grows in dry areas and is a little digestive with a bitter taste. It has a lot of body with little fragrance and contains about twice as much caffeine as Arabica. The robusta plants usually have a size that can reach up to 6 m, and according to ICO, their cultivation represents 42% of the world production; it is more resistant to attacks of parasites, diseases and high temperatures (hence its name).

Robusta grains are smaller than those of arabica. Depending on the variety of plant, the seed shape is round, oval or elliptical with sharp tips. Robusta varieties include Comilon, Kouilloi, Niaouli and Uganda.

The other variety is arabica coffee (*Coffea arabica*), a native species of Ethiopia, which also grows in other countries that are between 500 and 2400 m above sea level. This variety represents 58% of world production and has a caffeine concentration of up to 1.7%. The result of this is that arabica coffee origin is considered to be of much higher quality, not because of its much lower caffeine content, but because of its intrinsic, more aromatic and aromatic organoleptic qualities; therefore, it has more aroma and softness. The arabica seed is flattened and elongated and its green colour is more intense. Some subspecies of arabica are Moka, Maragogype, Bourbon, Mundo Novo, Caturra, Icatu, Catuai, Catimor, Creole, among others.

3. Methods of characterization of coffee

Many authors that carry out investigations with coffee use different methodologies to characterize. Proximate composition, reducing and total sugar contents are evaluated on the basis of standard methodologies [8]. The contents of cellulose, hemicellulose and lignin are determined by crude fibre analysis [9].

Sugars, ethanol, glycerol and volatile fatty acids are determined by HPLC with a refractive index detector [10]. The concentration of glucose, xylose, arabinose, mannose and galactose also can be determined by HPLC using a refractive index detector. Furfural and hydroxymethylfurfural (HMF) are determined by HPLC using a UV detector [11]. Minor volatile components are analysed in a GC-MS with capillary-coated column [12].

Important studies of gas chromatography coupled to mass spectrometry (GC-MS) have been performed to determine which and how many compounds are responsible for coffee aroma. Volatile compounds can be extracted by the simultaneous distillation and extraction (SDE) method and analysed in a GC-MS system. This tool allows determining the presence of compounds of the families of pyrazines, furan derivatives, ketones, pyrroles, acids, phenolic derivatives, pyridines, aldehydes and thiophenes [13] (Authors).

There are standards (ISO) for the quality of green coffee and its derivatives for domestic and international marketing; determination of the moisture content, routine method, olfactory and visual examination, determination of foreign matter and defects, analysis of grain size manual screening, determination of insect-damaged dams, sample preparation for sensory analysis, storage and transport, to name a few.

4. Coffee cherry components

The cherry coffee (*C. arabica*) structure consists of the outer skin (pericarp), pulp (exocarp), mucilage (mesocarp), parchment (endocarp), silverskin (perisperm) and coffee seed (endosperm) as shown in **Figure 2**.

4.1. Pericarp

The pericarp is composed of outer three layers of the fruit: the exocarp, mesocarp and endocarp (**Figure 3**).

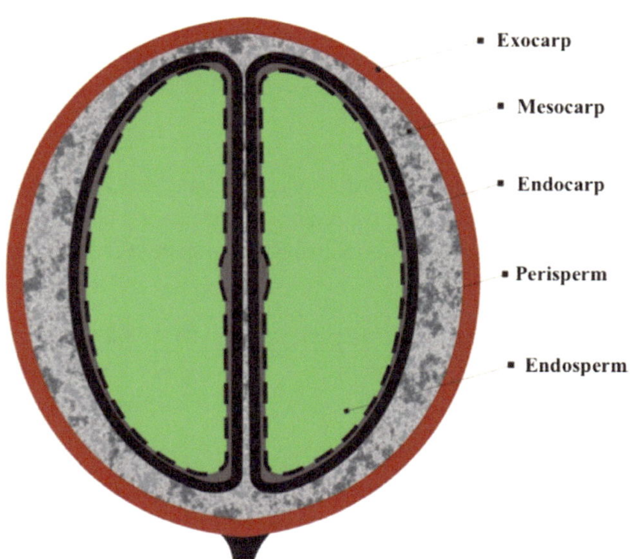

Figure 2. Coffee cherry structure (source: Authors).

Figure 3. Coffee cherry: exocarp, mesocarp bound endocarp (source: Authors).

4.1.1. Exocarp

The exocarp, also known as pulp, is the outer layer of cherry coffee. A single layer of cells of the compact parenchyma forms it. The colour of exocarp at the beginning of fruit growth is green due to the presence of chloroplasts that then disappear as the fruit matures. The colour of the ripening of the cherry depends on the variety of coffee that can be red or yellow. Red colour of the skin comes from anthocyanin pigments, whereas the yellow colour of the skin is due to luteins [14].

The coffee pulp when is discarded in the environment causes contamination. Due to this, many studies have been carried out to take advantage of it and reduce its toxic effect on the environmental process [15]. Among the ways of using it include silage for animal feed, coffee pulp cake, and juice treated with microorganisms for animal consumption [16]. The coffee pulp contains dry matter (92%), ethereal extract (2.6%), raw fibre (20.8%), crude protein (10.7%), ash (8.8%), nitrogen (49.2%), organic acids (12%), caffeine, trigonelline and tannins (1.8%) [17].

4.1.2. Mesocarp

The mesocarp or mucilage (a type of soluble and viscous fibre) is present in unripe coffee fruit. With maturation, pectolytic enzymes break down pectic chains, resulting in an insoluble hydrogel that is rich in sugars and pectins. Studies have shown that the mucilage/water ratio increases as altitude increases [14].

Coffee mucilage is a viscous liquid residue produced in the coffee industry that is disposed of without treatment directly into watercourses, causing serious contamination problems. The mucilage is composed of water (84.2%), protein (8.9%), reducing sugar (4.1%), pectates (0.91%) and ash (0.7%) [18].

Carbohydrates are the most important constituents in coffee mucilage. Also, syringaldehyde, which is produced by lignin hydrolysis, is found in low concentrations. Coffee mucilage (CM) contains several minerals; potassium is the most abundant element, followed by phosphorus, calcium, sulphur and magnesium (**Table 2**). Other compounds are also found, such as glycerol, caffeine, acetates, lactates, phenol, as well as 2,6 and 3,4-dimethoxyphenol.

Minerals	(mg/L)	Minerals	(mg/L)
Aluminium	0	Manganese	0.07
Arsenic	0.47	Molybdenum	0
Sulphur	30.19	Sodium	7.18
Boron	0.16	Nickel	0.01
Barium	0.02	Phosphorus	41.55
Beryllium	0	Lead	0
Calcium	37.08	Antimony	0
Cadmium	0	Selenium	0
Cobalt	0	Silicon	1.58
Chrome	0	Tin	0
Copper	2.45	Strontium	0.07
Iron	0.65	Thallium	0
Potassium	239.8	Vanadium	0
Lithium	0.01	Zinc	0.14
Magnesium	10.05		

Table 2. Minerals composition of coffee mucilage [10] (Authors).

4.1.3. Endocarp

The endocarp, or parchment, is the hard layer that surrounds the coffee seed. It consists of three to seven layers of sclerenchyma cells. These cells harden during the ripening of the coffee fruit, delimiting the size of the coffee seed [14].

In processing of coffee, parchment or husks are the major solid residues and it is estimated that for every kilogram of coffee seeds produced, approximately 1 kg of husks is generated. It is mainly composed of cellulose (40–49%), hemicellulose (25–32%), lignin (33–35%) and ash (0.5–1%) [19].

4.2. Seed

The perisperm, the endosperm and the embryo compose the coffee seed. The size of seeds vary according to the variety of coffee, usually the average can be between 10 mm long and 6 mm wide.

Quality of the coffee seed has a high and direct influence on the success of the crop and is directly dependent on viability, identity, health and appearance.

The health of the seed influences its germination, appearance and vigour, related to the health of the plants, depending on its management and the environmental conditions. The appearance of the seed has to do with its colour, and it must be a homogeneous amber yellow colour, without spots, without blows, without signs and symptoms of diseases, and without the remains of coffee by-products. Viability is the ability of the seed to germinate properly, giving rise to healthy and vigorous plants. To the genetic correspondence of plants to variety [20].

4.2.1. Perisperm

The perisperm or silverskin is the outer layer that surrounds the seed. It is formed from the nucellus (the central cell mass of the ovule's body). Usually some remnants of the silver skin remain in the seed, but when being roasted they are detached. The silverskin can be polished off the grain; however, this decreases the coffee flavour. Some authors claim they have proposed that the presence of a large amount of silver skin in the ground coffee is a sign that the coffee was cut before its ideal maturity to be processed [14]. The main component is cellulose and others.

4.2.2. Endosperm

The endosperm is the tissue produced inside the seed. The chemical content of the endosperm is very important as it precedes the taste and aroma of roasted coffee. Water-soluble compounds are caffeine, trigonelline, nicotinic acid (niacin), chlorogenic acids, monosaccharides, disaccharides, oligosaccharides, proteins, minerals and carboxylic acids [14]. Components insoluble in water include cellulose, polysaccharides, lignin and hemicellulose, as well as some proteins, minerals, vitamins and lipids (triglycerides and esters of diterpene alcohols and fatty acids). The most abundant amino acids are 17, and glutamic acid, aspartic acid, leucine and valine stand out among them.

5. Chemical composition of green coffee and roasted coffee

The green coffee seed is the fruit obtained from the trees of the genus *Coffea*; peeled, decaffeinated and ready for roasting, it is called raw coffee or gold coffee. Roasted coffee is the product obtained from the roasting of green coffee. The green coffee is roasted with heat at 180–230°C for 15–20 min leading to increase in size due to the production of carbon dioxide inside, which acts as a preservative until released by grinding.

Generally, raw green coffee contains water, protein, caffeine, lipids, soluble carbohydrates, insoluble polysaccharides, acids (soluble and non-volatile), trigonelline, amino acids and minerals (**Table 3**). Roasted coffee contains reducing sugars, caramelized sugars, insoluble polysaccharides, fibre, proteins, minerals, non-volatile acids (caffeic, chlorogenic, citric, malic, oxalic, quinic, tartaric), caffeine, lipids, trigonelline and ash, in which the main constituent elements are potassium, phosphorus and magnesium (**Table 4**).

The protein volatiles obtained by pyrolysis have some relevance in relation to the coffee flavour. Amino acids containing sulphur, methionine and cysteine have been identified along

Component	Arabica*	Constituents
Soluble carbohydrates	**9–12.5**	
Monosaccharide	0.2–0.5	
Oligosaccharide	6–9	Fructose, glucose, galactose, and arabinose (traces)
Polysaccharides	3–4	Sucrose (>90%), raffinose (0–0.9%), stachyose (0–0.13%), and glucose (0–2%)
Insoluble polysaccharides	46.53	
Hemicellulose	5–10	Polymers of galactose (65–75%), arabinose (25–30%), and mannose (0–10%)
Cellulose	41–43	
Volatile acids	0.1	
Non-volatile aliphatic acids	2–2.9	Citric acid, malic acid, and quinic acid
Chlorogenic acid	6.7–9.2	Mono-, dicaffeoyl- and feruloylquinic acid
Lignin	1–3	
Lipids	15–18	
Wax	0.2–0.3	
Oil	7.7–17.7	Main fatty acids: 16:0 and 18:2 (9,12)
N compounds	11–15	
Free amino acids	0.2–0.8	Main amino acids: Glu, Asp and Asp-NH_2
Proteins	8.5–12	
Caffeine	0.8–1.4	Traces of theobromine and theophylline
Trigonelline	0.6–1.2	
Minerals	3–5.4	

* Values in percent dry-weight basis.

Table 3. Chemical composition of green coffee [13].

with indole and tryptophan. Volatile matter includes numerous compounds such as acids, alcohols, aldehydes, diacetyl, furfural, hydrogen sulphide, ketones, mercaptans and phenols [21].

On the other hand, because of caffeine, drinking coffee can significantly affect the nervous system, cardiovascular, respiratory, among others. However, caffeine does not accumulate in the body, so its effects become short-lived. The body can become accustomed to caffeine and

Component	Average*
Cellulose	8.6
Hemicellulose	36.7
Xylose	0
Arabinose	1.7
Galactose	13.8
Mannose	21.2
Protein	10
Lipids	11–16
Ashes	1.6
No volatile acids	0.4
Soluble	24
Insoluble	4
Organic matter	90.5
Nitrogen	2.3
Carbon/nitrogen	22/1
Caffeine	1.2–2.4
Trigonelline	0.4
Protein	9
Minerals	4

* Values in percent dry-weight basis.

Table 4. Chemical composition of roasted coffee ground (seed) [24, 25].

make regular users less sensitive to its stimulating effects than others. Among all the effects of coffee, the best known is to be a stimulant to the nervous system. Consuming coffee can make one feel more awake, alert and able to concentrate. Caffeine has been shown to counteract fatigue and wake up the mood, but it can also cause anxiety, nervousness and irritability. In some people caffeine can delay sleep, but it all depends on how much has been consumed [22, 23].

Tables 3 and **4** show the chemical composition of green coffee and roasted coffee.

In samples of catuai (variety of *Arabica coffee*) roasted coffee, 111 compounds were isolated and identified, among which 7 pyrazines and 10 furan derivatives were outstanding. The pyrazines are important contributors in the coffee aroma, 2-ethyl-3,5-dimethyl-pyrazine was the most abundant followed by 2-ethyl-5-methylpyrazine in this sample.

Many furanic compounds are common in samples of roasted coffee. In samples of roasted bitter coffee, 119 isolates and 16 compounds were obtained. Of which 8 are pyrazines and 15 are furan derivatives. Of the pyrazines, the 2-ethyl-3,5-dimethyl-pyrazine was found in greater abundance followed by 2-ethyl-6-methylpyrazines. Similarly, in roasted catuai coffee, furanic derivatives such as 2-furancarboxaldehyde, 2-furanmentol and 5-methyl-2-furancarboxyaldehyde were found [13] (Authors).

In robusta coffee, 122 compounds were found and 120 compounds were identified. Of which only 18 are pyrazines and 11 are furanic derivatives. Among pyrazines, 3,5-dimethyl-2-propylpyrazine is the most abundant derivative in this coffee and compared to the other varieties is the only sample that is presented and is followed by 2,5-methyl-pyrazine.

2-furancarboxyaldehyde is one of the most abundant compounds found in all the varieties and provides a sweet aroma to the coffee and is a very penetrating caramel in the coffee variety of Kilimanjaro [26].

The 5-methyl-2-furancarboxyaldehyde and 2-furanmentol compounds were found in different concentrations in the three analysed varieties, of which the specific contribution they have in the coffee aroma has not yet been reported. But it belongs to the group of furanic derivatives that provide the note of roasted coffee. It is known that phenolic compounds are products of the thermal degradation of carbohydrates, chlorogenic acids and lignin substances [27].

6. Biotransformation of the main coffee components for biofuels production

Since more than 50% of the coffee fruit is not used for the production of commercially available green coffee, it is therefore discarded during processing. So far, most of the advances have been made in its use for industrial purposes other than the food industry, such as energy production, adsorption of compounds and the manufacture of industrial products such as particle boards, ethanol, gibberellic acid and α-amylase [28].

With 14% of a total of 18%, bioenergy is the largest source of renewable energy. In contrast to other sources of renewable energy, biomass can be transformed into solid, liquid and gaseous fuels. This is shifting from an unusual source of energy to an increasingly globalized market [29].

Bioethanol and biogas production by fermentation has received great importance in the last years due to its increase of the demand of fuels. Fermentation is one of the most important processes for agro-waste reuse producing yeast and clean fuels. This process does not require the use of toxic substances; this makes it an environmentally friendly process.

The outer layers of coffee cherry (pulp, skin, mucilage, etc.) are removed by various processes, including washing, drying and fermentation. Useless waste products, grains and coffee are classified. During these processes are generated different residues that being rich in sugar and compounds with functional properties do not receive adequate treatment and become sources of contamination of rivers and streams, mainly [30].

In order to carry out the extraction of the coffee components to be used as substrate to produce biofuels, among the different methods of extraction, the mechanical extraction of the coffee seeds reduces the amount of water used, and consequently allows the recovery of the fraction of mucilage, pulp and seed of coffee [10, 31].

Some applications of cherry coffee residues such as pulp, mucilage, parchment and coffee seed are processed through fermentation, anaerobic digestion and trans-esterification to produce chemicals (**Figure 4**).

6.1. Fermentation

Fermentation is a biological process in which complex molecules are degraded to transform them into simpler molecules, generating liquid products. Bioethanol is produced by alcoholic fermentation of sugars; this biofuel is considered as a good candidate to replace conventional fossil fuels. The advantages of biofuel over fossil fuels are that it is clean, renewable and fully combustible and generates less waste. However, the implementation of biofuel technology is intended to solve the energy problem that is presented, in order to reduce the current dependence on fossil fuels [32].

In the investigation of Pérez-Sariñana et al., a response surface design was performed to obtain the optimal operating conditions of fermentation system using coffee mucilage as a substrate and *Saccharomyces cerevisiae*, the aim being to obtain a maximum bioethanol production. The conditions were pH of 5.1, temperature of 32°C and initial sugar concentration of 61.8 g/L. With this, the estimated production of bioethanol was 15.02 g/L and the experimental production was 16.29 g/L ± 0.39 g/L. It was demonstrated that carbohydrates are the most important constituents in coffee mucilage; for samples analysed here, the sugar composition

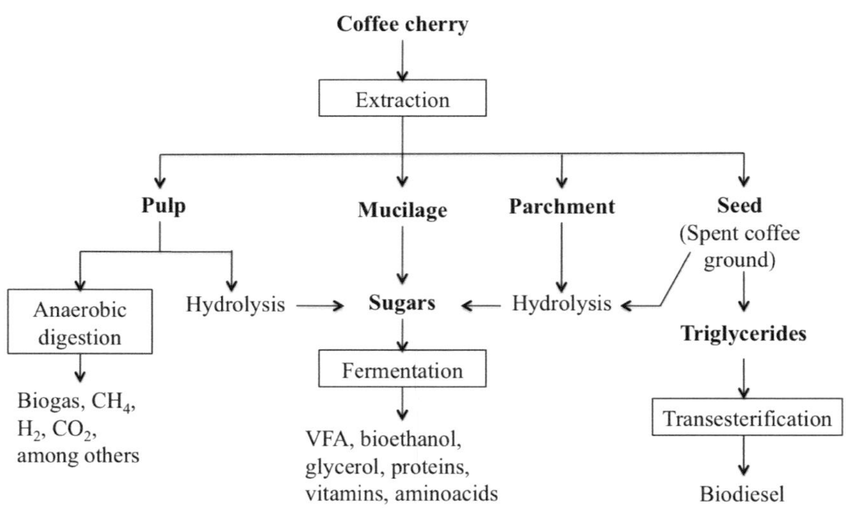

Figure 4. Applications for pulp, mucilage, parchment and seed of coffee using three processes—fermentation, anaerobic digestion and transesterification (Source: Authors).

in coffee mucilage was 37.67 g/L galactose, 35.65 g/L glucose, 1.06 g/L lactose and 0.1193 g/L proteins. The fermentation medium used to propagate the yeast was YPD (yeast extract, peptone and dextrose) agar, 1% yeast extract, 2% peptone, 2% glucose and 2% agar. The medium was pasteurized at 65°C for 30 and 20 min on ice and was supplemented with 0.5 g/L ammonium sulphate as a nitrogen source. Subsequently, the required volume was transferred to each serological bottle previously sterile. The strain was cultured in serological flasks, stirred at 200 rpm and 28°C [10] (Authors).

The use of the mucilage of coffee as a substrate by its chemical composition in sugars such as glucose and galactose allows or favours an adequate management of agro-industrial residues in the coffee-growing.

Harsono et al., carried out research on how to use coffee residues to produce value-added products and reduce the impacts of pollution on the environment, as well as to evaluate the bioethanol production potential (estimated optimum conditions) using *S. cerevisiae* yeast. They obtained a yield of 77.29% of bioethanol, which they consider may be a viable alternative for obtaining second generation bioethanol specifically in rural areas and for plantations of small coffee producers. They also assessed the cost of producing bioethanol that was evaluated from the processing of residual coffee [33].

Thnari et al. have used coffee residues for the potential they manifest with a dual-purpose in the production of ethanol and the preparation of activated carbon. A direct method of hydrolysis and direct fermentation is considered as the main option used in this study for the generation of ethanol fuel from biomass residues. Factors such as *S. cerevisiae* fillers, temperatures and substrate content were investigated to maximize ethanol yield. Coffee extract residue was also used to prepare activated carbons using chemical and physical activation methods. The effects of process parameters such as temperatures and acid concentrations were varied and determined in terms of yield, BET surface areas and porosity of the final product [34].

Other research points to the analysis of the carbohydrate content of coffee residues waste for fermentable sugars such as glucose, galactose and mannose, which can be fermented by *S. cerevisiae*. The rate of enzymatic conversion of coffee residues waste into fermentable sugars was 85.6%. The concentration of ethanol and yield (based on sugar content) after enzymatic hydrolysis, by simultaneous saccharification and by fermentation were 15.3 g/L and 87.2%, respectively [35].

6.2. Anaerobic digestion

Anaerobic digestion is a biological process in which organic matter is decomposed into different gaseous products by the action of a consortium of microorganisms.

In this context, biogas from waste will play a vital role in the future, as biogas is a versatile source of renewable energy that can be used to replace fossil fuels in energy and heat production, and can also be used as a gaseous fuel for vehicles. Biogas rich in methane (biomethane) can also replace natural gas and can also be used as raw material to produce chemicals and materials. The production of biogas through anaerobic digestion through the use of locally available resources offers significant advantages over other forms of bioenergy production. It has been evaluated as one of the most energy efficient and environmentally beneficial technologies [36]. It can drastically reduce GHG emissions compared to fossil fuels. Another advantage is that it is produced as digestate residue that is an improved fertilizer for crops that can substitute mineral fertilizer.

In the work of Pérez-Sariñana et al., pulper with desmucilating was used, a litre of water was added to a kilogram of coffee cherry and the coffee mucilage extracted was stored in bottles at −20°C to prevent degradation due to the sugar content it has. The concentrations of sugars in the experimental design were (72, 65, 50, 35 and 27 g/L). Optimal conditions for the methane production from coffee mucilage using methanogenic sludge as an inoculum and buffer solution (minerals) were estimated by the software as pH 8.2, temperature 37°C and sugar concentration of 27 g/L. The experimental optimum conditions for the production of methane from coffee mucilage were identified, which were pH 8.2, temperature 37°C, sugar concentration of 25.5 g/L and 313 mL methane .

On the other hand, Corro et al. mentioned that biogas could be produced by co-digestion of coffee pulp and cow manure under solar radiation. They reported that the methane content in the biogas reached 50% of the yield. This content increased to 60% and remained almost constant for at least 8 months of additional digestion. By means of gas spectroscopy analysis, more than 70 chemical compounds were found in the biogas generated after 4 months of co-digestion [37].

Hernández et al. used the coffee mucilage as a substrate for the production of hydrogen. The study evaluated three proportions of mucilage manure by performing a co-digestion and also increased the organic load to improve hydrogen production. The average rate of hydrogen production reached 7.6 NLH_2/Ld of coffee mucilage (parameters as the hydrogen production rate), indicating a high potential for hydrogen compared to substrates such as palm oil and wheat starch [38].

Luongo et al. (2015) indicated that methane-specific production reached 0.15 $NLCH_4$/g TVS using glucose as the most readily biodegradable carbon source, and a material rich in lignocelluloses (coffee seed skin). The application of multiple anaerobic digestion extracts more energy from organic waste [39].

A continuous flow stirred tank reactor was started for the treatment of coffee residues at thermophilic temperatures and long-term operation. In this experiment, the reactor was fed a substrate mixture (total solids of about 70 g/L) of ground coffee, coffee wastewater, milk waste and municipal sludge and was run at 55°C for 225 days. They show that the effectiveness of the complete parameters (total volatile fatty acids, propionic acid, intermediate alkalinity/partial alkalinity, intermediate alkalinity/total alkalinity and CH_4 content) controlled the thermophilic system [40].

6.3. Trans-esterification

Trans-esterification is a chemical reaction in which there is an exchange of the alkoxy group of an alcohol; the glycerol contained in the oils is replaced by an alcohol in the presence of a catalyst.

Biodiesel from vegetable oils, animal fats or other materials is an alternative to petroleum diesel for use in compression ignition engines. The composition and quality of biodiesel depends greatly on the composition of the raw material used. In the trans-esterification process for biodiesel, monoalkyl esters are produced from a glycerol-containing vegetable oil of long chain fatty acids with a low molecular weight alcohol (methanol) [41].

Used or spent coffee seeds are currently being used to turn them into biodiesel. This process produces 10–15% oil depending on the species of coffee (Arabica or Robusta). Kondamudi et al. carried out research on coffee-derived biodiesel where it projects that 340 million gallons of biodiesel can be produced from coffee residues around the world [42].

Rocha et al. presented a study of ultrasonic-assisted extraction; they used solid residues from the coffee process as a substrate for the production of oils in order to produce biodiesel and ethanol. The process for producing biodiesel showed a yield of 97% in methyl esters of fatty acids. And the highest glucose yield (192 mg $gSCG^{-1}$) was obtained by hydrolysis with 0.4 mol/L sulphuric acid at 121°C for 15 min [43].

7. Conclusions

In this chapter, the characteristics of coffee from the production, main varieties, characterization methods, components of cherry coffee and residues that were discarded after the extraction of the coffee bean were presented. These residues are used to produce energy as it is raw material in the form of biomass; the products that can be obtained after being transformed into energy with fermentation processes and anaerobic digestion are liquid and gaseous biofuels (bioethanol, biodiesel and biogas).

The agro-industrial residues are generated by the fruit and vegetable sector primary activity that shows the need to propose alternatives to use of the residues with a very important potential to be used as a source of chemical components that can be used for the production of biofuels. Their energy potential allows it to know their usefulness in the production of bioethanol, biogas and biodiesel, as well as the alternatives of using the by-products of the processes elaborated in order to avoid the generation of residues that contaminate the soil, water or the air.

Most of the research work has been done at laboratory scale and pilot scale; the next step would be to assess the technical feasibility and energy efficiency of these processes to carry out scaling at the industrial level.

Acknowledgements

The authors thank the Universidad Politécnica de Chiapas and its Rector M.S. Navor Francisco Ballinas Morales for the support provided for the development of this work.

Appendices and nomenclatures

GC-MS	gas chromatography-mass spectrometry
HPLC	high performance liquid chromatography
ISO	International Organization for Standardization
CM	coffee mucilage
VFA	volatile fatty acids
CH_4	methane
H_2	hydrogen
CO_2	carbon dioxide
GHG	greenhouse gas
g	gram
L	litre
d	day
SCG	spent coffee grounds
H_2	hydrogen
CH_4	methane
NL H_2	parameters as the hydrogen production rate
NL CH_4	parameters as the methane production rate
TVS	total volatile solids
BET	Brunauer-Emmett-Teller
SDE	simultaneous distillation and extraction

Author details

Bianca Yadira Pérez-Sariñana* and Sergio Saldaña-Trinidad

*Address all correspondence to: bperez@ie.upchiapas.edu.mx

Polytechnic University of Chiapas, Suchiapa, Chiapas, Mexico

References

[1] Balat M, Balat H. Recent trends in global production and utilization of bioethanol fuel. Applied Energy. 2009;**86**:2273–2282

[2] Trade Statistics Tables Production by Exporting Countries [Internet]. Available from: http://www.icg/trade_statistics.asp?section=Statistics

[3] International Energy Agency (IEA). World Energy Outlook 2016—Executive Summary. Organisation for Economic Co-operation and Development/International Energy Agency Editor Publishing; 2016. pp. 1–13. Available from: http://www.iea.org/publications/freepublications/publication/WorldEnergyOutlook2016ExecutiveSummaryEnglish.pdf [Accessed: 14 Feburary, 2017]

[4] Renewable Energy Policy Network (REN21). Renewables-2017: Global Status Report. Paris and Washington, DC: REN21 and Worldwatch Institute; 2016. pp. 6–11. Available from: http://www.ren21.net/wp-content/uploads/2016/06/GSR_2016_KeyFindings_SPANISH.pdf [Accessed: 14 Feburary 2017]

[5] Rocha MH, Capaz RS, Lora EES, Nogueira LAH, Leme MMV, Renó MLG, Olmo OA. Life cycle assessment (LCA) for biofuels in Brazilian conditions: A meta analysis. Renewable and Sustainable Energy Reviews. 2014;**37**:435–459

[6] Dutta, K., Achlesh D, Jih-Gaw L. Evolution retrospective for alternative fuels: First to fourth generation. Renewable Energy. 2014;**69**:114–122

[7] Carneiro ML, Pradelle F, Braga SL, Gomes MSP, Martins AR, Turkovics F, Pradelle RNC. Potential of biofuels from algae: Comparison with fossil fuels, ethanol and biodiesel in Europe and Brazil through life cycle assessment (LCA). Renewable and Sustainable Energy Reviews. 2017;**73**:632–653

[8] Official methods of analysis of the Association of Official Analytical Chemists. William Horwitz, editor. 16th ed. Washington; 1995

[9] Gouvea, BM, Torres C, Franca S, Oliveira LS, Oliveira ES. Feasibility of ethanol production from coffee husks. Biotechnology Letters. 2009;**31**:135–139

[10] Pérez-Sariñana BY, DeLeón-Rodríguez A, Saldaña-Trinidad S, Sebastian PJ. Optimization of bioethanol production from coffee mucilage. Bioresources. 2015;**10**:4326–4338

[11] Mussatto SI, Machado EMS, Carneiro LM, Teixeira JA. Sugars metabolism and ethanol production by different yeast strains from coffee industry wastes hydrolysates. Applied Energy. 2012;**92**:763-768

[12] Sampaio A, Dragone G, Vilanova M, Oliveira JM, Teixeira JA, Mussatto SI. Production, chemical characterization, and sensory profile of a a novel spirit elaborated from spent coffee ground. LWT-Food Science and Technology. 2013;**54**:557–563

[13] Saldaña-Trinidad S, Hernández HDM, López PMG. Análisis de café chiapanecoporcromatografía de gases acopladosaespectrometría de masas. In: III SimposioInternacional de Ciencia y Tecnología de Alimentos. IV CongresoInternacional de Evaluaciones Sensorial; 30–31 de agosto y 1 de septiembre; Villahermosa, Tabasco. México; 2006. pp. 1–6

[14] Castro R, Marraccini P. Cytology, biochemistry and molecular changes during coffee fruit development. Brazilian Journal of Plant Physiology. 2006;**18**:175–199

[15] Ramirez J. Coffee pulp is a by-product, not a waste. Tea Coffee Trade Journal. 1998;**170**: 116–123

[16] Ferrer J, Páez G, Chirino M, Mármol Z . Ensilaje de la pulpa de café. Revista de la Facultad de Agronomía. 1995;**12**:417–428

[17] Bressani R. Potential uses of coffee berry by-products. In: Braham JE, Bressani R, editors. Coffee Pulp: Composition, Technology and Utilization. Guatemala City: Institute of Nutrition of Central America and Panama; 1987. pp. 17–25

[18] Belitz HD, Grosch W, Schieberle P. Coffee, tea, cocoa. In: Belitz HD, Grosch W, Schieberle P, editors. Food Chemistry. 4th ed. Leipzig: Springer; 2009. pp. 938–951

[19] Bekalo SA, Reinhardt HW. Fibers of coffee husk and hulls for the production of particleboard. Materials and Structures. 2010;**43**:1049–1060

[20] Cenicafe, Grow coffee. Quality Criteria of the Coffee Seed [Internet]. Available from: http://www.cenicafe.org/es/index.php/cultivemos_cafe/semilla

[21] Kirk R, Sawyer R, Egan H. Pearson's composition and analysis of foods. In: Beverages and Chocolate. 9th ed. Longman Scientific and Technical; Harlow, Essex, UK;1991. pp. 398–399

[22] Caffeine. Summary of the Physiological Effects of Caffeine [Internet]. Available from: http://www.ico.org/ES/caffeine_c.asp

[23] Aarthi L, Persad B. Energy drinks and the neurophysiological impact of caffeine. Frontiers in Neuroscience. 2011;**5**:116

[24] Mussatto SI, Carneiro LM, Silva JPA, Roberto IC, Teixeira JAA. Study on chemical constituents and sugars extraction from spent coffee grounds. Carbohydrate Polymers. 2011;**83**:368–374

[25] ABNT AssociaçãoBrasileira de NormasTécnicas. Residuos Sólidos Classificação. Rio de Janeiro, Brazil: ABNT 10.004; 1987

[26] Murota A. Canonical discrement analysis applied to the head space GC profiles of coffee cultivars. Bioscience, Biotechnology and Biochemistry. 1993;**57**:1043–1048

[27] Yamanishi T. Flavor research–recent advances. In: Teranishi R, Flath RA, Sugisawa H, editors. Tea, Cocoa and Other Beverages. New York: Dekker, Inc.; 1981. pp. 278–287

[28] Esquivel P, Jimenez VM. Functional properties of coffee and coffee by-products. Food Research International. 2012;**46**:488–495

[29] World Energy Council. World Energy Resources [Internet]. 2016. Available from: www.worldenergy.org [Accessed: 25 January, 2017]

[30] Avallone S, Guiraud J, Guyot B, Olguin E, Brillouet M. Polysaccharide constituents of coffee-bean mucilage. Jorunal of Food Science. 2000;**65**:1308–1311

[31] Pérez-Sariñana BY, Saldaña-Trinidad S, Guerrero-Fajardo CA, Sebastian PJ. Ethanol production by fermentation from waste coffee mucilage. Indian Journal of Applied Research. 2015;**5**:197–199

[32] Pérez-Sariñana BY, Saldaña-Trinidad S, Fernando SEL, Sebastian PJ, Eapen D. Bioethanol production from coffee mucilage. Energy Procedia. 2014;**57**:950–956

[33] Harsono SS, Fauzi M, Purwono GS, Soemarno D. Second generation bioethanol from Arabica coffee waste processing at smallholder plantation in Ijen Plateau Region of East Java. In: HK-ICONS 2014 2nd Humboldt Kolleg in Conjuction with International Conference on Natural Science; Procedia Chemistry; 2015. pp. 408–413

[34] Thnari NF, Aznar JS, Kiros Y. Coffee extract residue for production of ethanol and activated carbons. Journal of Cleaner Production. 2015;**91**:64–70

[35] Choi IS, Wi SG, Kim SB, Bae HJ. Conversion of coffee residue waste into bioethanol with using popping pretreatment. BioresourceTechonolgy. 2012;**125**:132–137

[36] Fehrenbach H, Giegrich J, Sayer U, Gretz M, Lanje K, Schmitz J. Kriterieneinernachhaltigen Bioenergienutzungimglobalen Maßstab. UBA-Forschungsbericht. 2008;**206**:41–112

[37] Corro G, Paniagua L, Pal U, Bañuelos F, Rosas M. Generation of biogas from coffee-pulp and cow-dung co-digestion: Infrared studies of postcombustion emissions. Energy Conversion and Management. 2013;**74**:471–481

[38] Hernández MA, Rodríguez SM, Yves A. Use of coffee mucilage as a new substrate for hydrogen production in anaerobic co-digestion with swine manure. Bioresource Technology. 2014;**168**:112–118

[39] Luongo MAC, Bernardi M, Fino D, Ruggeri B. Multistep anaerobic digestion (MAD) as a tool to increase energy production via H_2 + CH_4. International Journal of Hydrogen Energy. 2015;**40**:5050–5061

[40] Shofie M, Qiao W, Li Q, Takayanagi K, Li Y. Comprehensive monitoring and management of a long-term thermophilic CSTR treating coffee grounds, coffee liquid, milk waste, and municipal sludge. Bioresource Technology. 2015;**192**:202–211

[41] Gerhard K, Razon LF. Biodiesel fuels. Progress in Energy and Combustion. 2017;**58**:36–59

[42] Kondamudi N, Mohapatra SK, Misra M. Spent coffee grounds as a versatile source of green energy. Journal of Agricultural and Food Chemistry. 2008;**56**:11757–11760

[43] Rocha MV, De Matos LJ, Lima LP, Figueiredo PM, Lucena IL, Fernandes FA, Gonçalves LR. Ultrasound-assisted production of biodiesel and ethanol from spent coffee grounds. Bioresource Technology. 2014;**167**:343–348